SURFACE ANALYSIS BY ELECTRON SPECTROSCOPY
Measurement and Interpretation

UPDATES IN APPLIED PHYSICS AND ELECTRICAL TECHNOLOGY

Series Editor: P. J. Dobson
Oxford University
Oxford, England

CRYSTAL GROWTH: Principles and Progress
A. W. Vere

DRY ETCHING FOR VLSI
A. J. van Roosmalen, J. A. G. Baggerman, and S. J. H. Brader

OPTICAL FIBRES AND SOURCES FOR COMMUNICATIONS
M. J. Adams and I. D. Henning

PHOTODETECTORS: An Introduction to Current Technology
P. N. J. Dennis

SURFACE ANALYSIS BY ELECTRON SPECTROSCOPY: Measurement and Interpretation
Graham C. Smith

SURFACE ANALYSIS BY ELECTRON SPECTROSCOPY
Measurement and Interpretation

Graham C. Smith

Shell Research Ltd.
Chester, England

PLENUM PRESS • NEW YORK AND LONDON

Library of Congress Cataloging-in-Publication Data

Smith, Graham C.
    Surface analysis by electron spectroscopy : measurement and
interpretation / Graham C. Smith.
        p.   cm. -- (Updates in applied physics and electrical
technology)
    Includes bibliographical references and index.
    ISBN 0-306-44806-8
    1. Surfaces (Technology)--Analysis. 2. Surface chemistry.
3. Auger effect. 4. X-ray spectroscopy.   I. Title. II. Series.
TA418.7.S57 1994
537.5'352--dc20   94-35640
                                                               CIP

ISBN 0-306-44806-8

©1994 Plenum Press, New York
A Division of Plenum Publishing Corporation
233 Spring Street, New York, N.Y. 10013

To Ruth

This book is the fifth in a series of scientific textbooks designed to cover advances in selected research fields from a basic and general viewpoint. The reader is taken carefully but rapidly through the introductory material in order that the significance of recent developments can be understood with only limited initial knowledge. The inclusion in the Appendix of the abstracts of many of the more important papers in the field provides further assistance for the non-specialist, and acts as a springboard to supplementary reading for those who wish to consult the original literature.

Surface analysis has been the subject of numerous books and review articles, and the fundamental scientific principles of the more popular techniques are now reasonably well established. This book is concerned with the very powerful techniques of Auger electron and X-ray photoelectron spectroscopy (AES and XPS), with an emphasis on how they may be performed as part of a modern analytical facility. Since the development of AES and XPS in the late 1960s and early 1970s there have been great strides forward in the sensitivities and resolutions of the instrumentation. Simultaneously, these spectroscopies have undergone a veritable explosion, both in their acceptance alongside more routine analytical techniques and in the range of problems and materials to which they are applied. As a result, many researchers in industry and in academia now come into contact with AES and XPS not as specialists, but as users. A user of a technique needs an appreciation of its power and its limitations, together with an understanding of the instrumentation and an idea of the directions in which the field is developing. This book addresses these needs.

Although many aspects of AES and XPS are well established, and the techniques are in routine operation in hundreds of laboratories worldwide, there are still areas of controversy. The reader is led through sufficient material to be able to appreciate, for example, the importance of the spectral background in quantification, and why this issue is not yet completely resolved. Similarly, the ultimate resolutions and sensitivities of the techniques are not yet totally clear, and certain aspects of the data processing require further research before they attain the limits of reliability expected routinely in more conventional analyses.

The overall objectives of the book are to enable the reader with possibly a physical, chemical or materials sciences background to understand the principles of the techniques, to be able to assess whether they are the correct techniques for a particular application, to appreciate the design and operation of the instrumentation, and to know the issues that may be treated routinely and separate them from those aspects where particularly careful thought is needed. Finally, no analytical technique should operate in isolation, and it will frequently be necessary to compare and contrast the results of surface analysis by AES or XPS with conclusions drawn from other methods. To this end, a short chapter comparing the electron spectroscopies with complementary analytical techniques is included.

CONTENTS

CHAPTER 1

INTRODUCTION

Electron spectroscopy is applied to the analysis of surfaces in several thousand laboratories worldwide. It is a pervasive activity, spread across a wide range of scientific and technological subject areas, and has experienced a period of rapid and continued growth since its emergence from pure research in the late 1960s and early 1970s. There is some evidence to suggest that the principle electron spectroscopic techniques of surface analysis, namely X-ray photoelectron spectroscopy (XPS) and Auger electron spectroscopy (AES), have now reached a stage of maturity in their development at which they can be applied in a routine manner to solve practical problems alongside the more conventional techniques found in the well-equipped analytical laboratory. Applications occur in all the technologically advanced industries.

In the context of this book, surface analysis is concerned with the determination of the composition, chemical structure and local arrangement of the atoms that make up the top few nanometres of, usually, a solid specimen.

All the various surface analysis applications have differing scientific and instrumental requirements for their solution. However, any method that is proposed for the analysis of surfaces must satisfy certain basic criteria. It must be sensitive to the top few atomic layers of the sample under investigation. This means a depth sensitivity of the order of 1 nm or better is called for. The technique must be able to distinguish between different elements and, preferably, between different chemical states of the same element. As well as compositional information, the ability to provide data on structure and local atomic arrangement is desirable. Spatial resolution is obviously of great importance for any microanalytical technique and, of course, better spatial resolution is always required. For general work, something perhaps in the range around 1 μm is satisfactory. The techniques must exhibit sensitivity; that is, the ability to detect a low concentration of one material spread throughout a matrix of another. A general requirement of all analytical techniques which must be met is that of rapid turn-around time, particularly in a quality-control or trouble-shooting environment. There would be no point in calling in surface analysis to help solve a problem causing the shut-down of a electronics production facility, for example, if results were going to be obtained in a matter of weeks rather than hours. Finally, the data produced must be capable of quantitative interpretation. There may be some advantage in rapid identification of the presence of an element in a particular sample; this will always be followed by the need to determine what state it is in and how much is present.

These requirements in total are very demanding and cannot be met by any one technique. There are a large number of possible methods of surface analysis of which those based on electron spectroscopy are the most well established. This book focusses on the two most popular electron spectroscopies used for surface analysis; namely Auger electron spectroscopy (AES) and X-ray photoelectron spectroscopy (XPS). AES can operate with spatial resolution in the 50 to 100 nm range with sensitivity down to around 0.1% of a single atom layer. X-ray photoelectron spectroscopy (XPS) has rather poorer spatial resolution generally of the order of typically 100 μm in conventional instruments, although current developments are pushing it nearer to the 1 μm range. Sensitivity is similar to AES, but XPS readily lends itself to the identification of chemical states in addition to element types. Both AES and XPS may be made quantitative with reasonably good precision; if accuracy is required this is possible but a great deal of care must be taken. Either AES or XPS may be combined with in-situ erosion of the sample by an ion beam to produce profiles of concentration variations to depths of perhaps 0.5 to 1.0 μm from the sample surface. Although both AES and XPS are usually thought of as techniques for the examination of solids, this is not always the case, and progress has been made with XPS in particular in the analysis of liquid surfaces (Siegbahn et al, 1981; Baschenko et al, 1993).

The physical basis of these surface spectroscopies and the principles of spectral interpretation are described in the following chapter. Chapter 3 gives details of the instrumental requirements of the techniques, together with general information on spatial resolution, and methods required to avoid sample radiation damage and electrostatic charge build-up. The important subject of data reduction by computer-based methods is covered in Chapter 4. Quantification of the spectra is treated in Chapter 5. and the determination of sample structure using either argon-ion sputtering or non-destructive techniques is discussed in Chapter 6. The objective, in each case, is to provide an overview of the current state of the art and to introduce key references to the literature where further details on specific points may be followed up.

CHAPTER 2

SURFACE ANALYSIS BY ELECTRON SPECTROSCOPY

2.1 Surface Sensitivity

In AES and XPS, electrons are emitted from the sample as a consequence
of electron or X-ray irradiation respectively, and are subsequently energy-
analysed and detected. For electrons in the energy range 100 to 1000
electron volts (eV), the distance that may be travelled before undergoing an
inelastic collision, known as the inelastic mean free path, may be typically
of the order of 2 - 3 nm. This distance corresponds to perhaps 10 atom
layers in most materials, and it is this that gives the techniques their
surface specificity. Experimentally, this inelastic mean free path is very
difficult to measure and, in practice, a parameter known as the attenuation
length, which also includes the effect of elastic scattering, is determined
instead. Figure 2.1, from the work of Seah and Dench (1979), shows a compilation of measured attenuation length data for elements. These data are
primarily derived from thin overlayer experiments in which the structure of
the overlayers was usually not well characterised, with the result that the
average of the compilation is systematically low. Nevertheless, a broad
minimum in the energy range of interest is seen which rises at both high and
low energies. More recent estimations of the attenuation lengths of Auger
electrons and X-ray photoelectrons are discussed in Chapter 5.

2.2 X-ray Photoelectron Spectroscopy

XPS consists, in principle, of the application of energy analysis to the
electrons emitted from a surface illuminated by X-rays and exhibiting the
photoelectric effect. XPS as it is used today is a direct result of
pioneering work by Siegbahn and his group at Uppsala University in Sweden
(Siegbahn et al, 1969).

The physical basis of the XPS technique is shown in Figure 2.2. The
energy carried by an incoming X-ray photon is absorbed by the target atom,
raising it into an excited state from which it relaxes by the emission of a
photoelectron. Photoelectrons are emitted from all energy levels of the
target atom and hence the electron energy spectrum is characteristic of the
emitting atom type, and may be thought of as its XPS fingerprint. Lines in
the spectrum are labelled according to the energy level from which they
originate. For example, in the experimental XPS spectrum for copper shown in
Figure 2.3, lines are seen due to emission from the 2s, 2p, 3s levels and so

on. The energy scale is labelled here as binding energy, in electron volts,
starting with zero at the Fermi level. XPS spectra may also be labelled in
terms of the kinetic energy of the emitted electrons, where the kinetic
energy is obtained by subtracting the binding energy of the level of interest
and the work function of the target material from the X-ray photon energy.
Electron spectrometers function by determining the kinetic energies of
electron emitted from the sample surface. However, interpretation of the
spectra is facilitated by their presentation in terms of binding energy, with
the instrument's data system performing the conversion automatically.

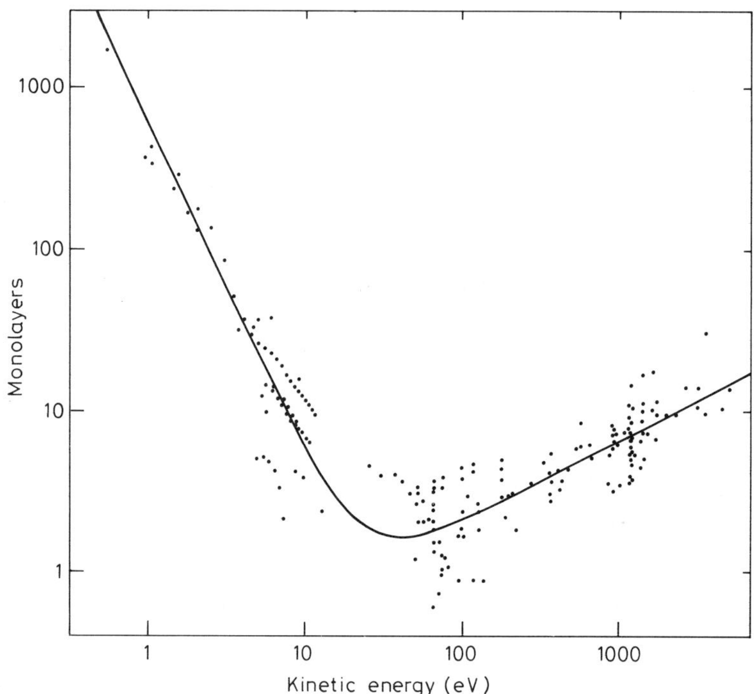

Figure 2.1 A compilation of experimentally-determined attenuation length
data, together with a best parametric fit to give the "universal
curve" of Seah and Dench (1979).
(reproduced with permission, John Wiley and Sons Ltd.)

    The intensities of the various photoelectron lines are important for
quantification and depend upon the amount of material present, the cross-
section for photoemission at the X-ray energy concerned, and various instrumental and geometrical factors. The measurement and calculation of XPS
intensities are discussed in Chapter 5. The energy widths of the peaks are
governed by the intrinsic widths of the energy levels involved and their
lifetimes in the excited state, coupled with instrumental broadening
resulting from the finite energy resolution of the electron energy analyser
and the natural width of the exciting X-ray line.

    The precise energy and shape of a particular line in an XPS spectrum is
a function not only of the emitting element but also of its local environment
and chemical state. Any effect that may cause a perturbation of the energy
levels of atoms in the surface region of the target will cause a concomitant
variation in the XPS spectrum. This gives rise to the chemical shifts in

energy that are observed in XPS spectra. These are extremely important for
spectral interpretation and are discussed in section 2.3 below. The so-
called surface core level shifts are a special case of this effect. Energy
shifts to lower binding energies are observed for emission from the energy
levels in atoms situated in the surface region of a crystalline solid. They
originate from the reduced coordination number of the atom concerned, giving
a consequently less tightly-bound energy level structure. Surface core level
shifts are usually only observed on particularly clean and well prepared
surfaces, and are not often encountered in practice.

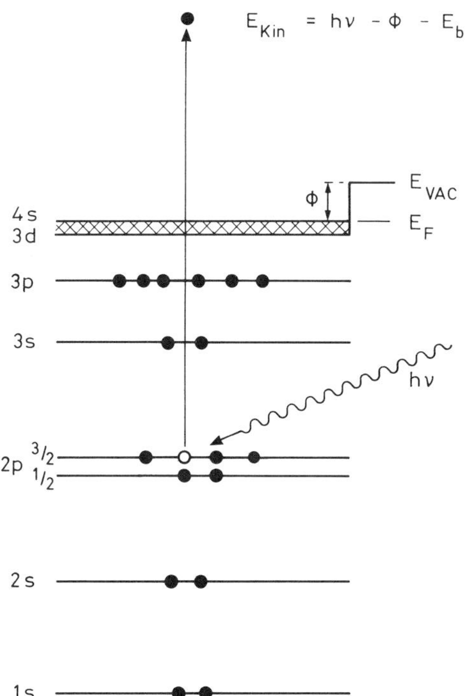

Figure 2.2 The electron transition involved in the photoemission of a $2p_{3/2}$
electron from copper.

      In addition to the characteristic peaks, the XPS spectrum in Figure 2.3
shows features due to the Auger effect, which will be described in the
following section, and both the XPS and Auger peaks are superimposed upon a
background. This background results from photoelectrons produced within the
target material that have been the subject of one or more inelastic scattering processes before emission from the surface. The background rises on the
low kinetic energy side of each peak, as a result of each individual photoemission line acting as a source of additional electrons which may subsequently undergo scattering events. Indeed, as we have already seen, it is the
presence of strong inelastic scattering that gives rise to the surface
sensitivity of the technique.

      XPS provides element and chemical state identification for atoms located
within the top few atomic layers of the sample under investigation. However,
under certain conditions, a degree of structural information may also be
obtained. The shape of the spectral background itself contains information
concerning the depth distribution of the emitting species. Furthermore, if

measures are taken to restrict the angle of emission of the electrons detected, then data concerning the depth distributions may be collected by comparing spectra taken near normal emission with those where emission is near glancing. As a result of the finite attenuation length of the photo-electrons, the grazing emission spectrum will show an increase in the intensities of lines due to elements distributed primarily in the outermost atomic layers. Similarly the normal-emission spectra have a relatively enhanced contribution from components below this outermost layer. The energy dependence of the attenuation lengths may also be employed to give in-depth information. Such methods are discussed further in Chapter 6.

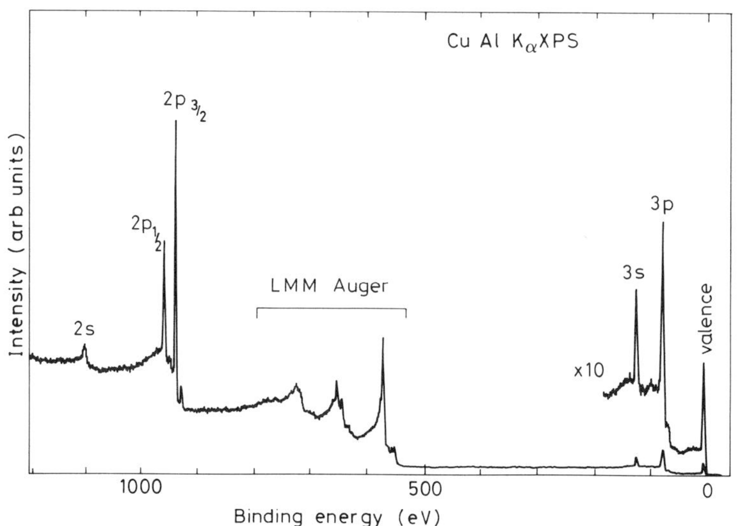

Figure 2.3 The X-ray photoelectron spectrum of a clean copper surface, excited by aluminium X-rays

In summary, XPS is an extremely powerful method of surface analysis which is frequently applied to a variety of technological problems, particularly in the fields of catalysis and polymer technology. However, in conventional use it has poor spatial resolution compared to AES, and is limited to samples with lateral homogeniety. Developments towards higher spatial resolution are discussed in Chapter 3. The X-ray beam is less damaging to the specimen surface than the electron beam used in AES, therefore XPS is the favoured technique for the analysis of the surface chemistry of radiation sensitive materials. The accumulation of electric charge during the analysis of insulating samples is also less of a problem in XPS than AES, as the positive sample charge formed because of the emission of electrons is readily neutralised by either a low energy electron beam, or by the stray secondary electrons often generated by the X-ray source.

2.3 Auger Electron Spectroscopy

In Auger electron spectroscopy (AES), electrons are detected after emission from the sample as the result of non-radiative decay of an excited atom in the surface region. The effect is named after the French physicist who first described the process involved (Auger, 1925). An inter-atomic process resulting in the production of an Auger electron is shown in Figure 2.4. The atom is raised into an excited state by the creation of a core

hole, as a result of an interaction with either an incident X-ray photon or
high energy electron. An electron then falls in energy from a higher level
to fill this core hole, and the excess energy is carried away as the kinetic
energy of a further electron which is emitted from the atom. X-ray nomenclature is used for the energy levels involved, and the Auger electron is
described as originating from, for example, an ABC Auger transition where A
is the level of the original core hole, B is the level from which the core
hole was filled and C is the level from which the Auger electron was emitted.
Thus, in the example of copper shown in Figure 2.4 the Auger transition
illustrated is described as $L_3M_1M_{2,3}$.

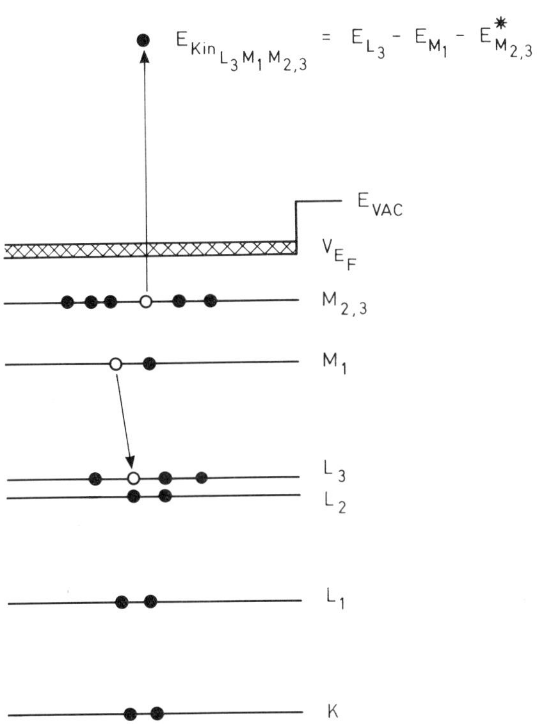

Figure 2.4 The electron transitions involved in the emission of an LMM Auger
electron from copper.

It is customary to use electron beam excitation in AES, although any
incident radiation capable of core-level ionisation can give rise to Auger
transitions. This accounts for the presence of the Auger peaks in the XPS
spectrum shown in Figure 2.3. Auger peaks may be distinguished from the
photoelectron peaks in XPS spectra by recording the spectrum using two
different photon energies, in which case the Auger peaks will remain at the
same kinetic energy in the spectrum whereas the photoelectron peaks will
shift according to the change in photon energy, as their binding energies
referenced to the Fermi level must stay constant.

The use of the electron beam as the primary method of excitation in AES
gives the technique one of its most powerful attributes, namely high spatial
resolution. The beam can be scanned, as in the scanning electron microscope,
and, by tuning the analyser to detect Auger electrons from the elements of
interest, compositional maps of surface concentration can be built up.

Currently, the state of the art limit of resolution is in the range 50 - 100 nm, although much useful work is carried out with beam diameters around 1 μm. This is discussed further in Chapter 3.

Figure 2.5 shows the Auger spectrum of copper, excited with a 5 keV electron beam. The direct spectrum is shown in Figure 2.5(a) and the differentiated spectrum in Figure 2.5(b). Both methods of data presentation have their advocates; their relative advantages when it comes to quantification of the spectra are discussed in Chapter 5.

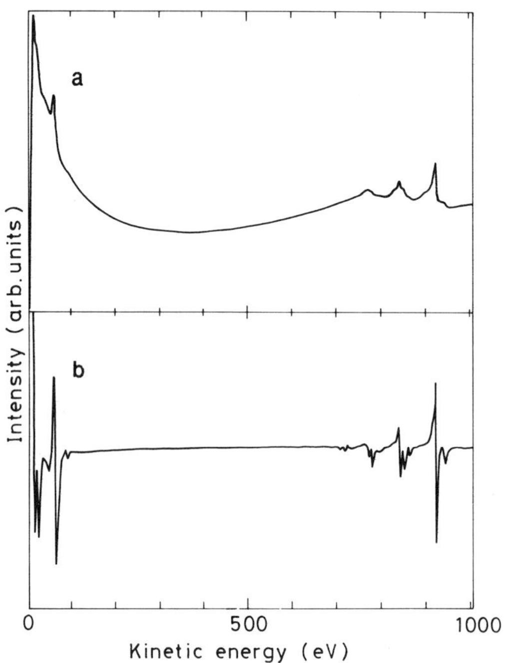

Figure 2.5 Auger spectra excited from a clean copper surface using a 5 keV electron beam. The upper curve (a) shows the direct spectrum and the lower curve (b) shows the derivative spectrum.

The calculation of the energies of lines in the Auger electron spectrum is complicated by the fact that emission occurs from an atom in an excited state and, consequently, the energies of the levels involved are difficult to define precisely. Ignoring this problem, in principle the kinetic energy of an electron resulting from an ABC Auger transition in a particular atom would be given by

$$E_{ABC} = E_A - E_B - E_C \qquad\qquad [2.1]$$

In the early years of AES, a good deal of effort was invested in determining approximate methods of estimating the modifications to E_B and E_C arising from the excited nature of the emitting atom in order to assist in the calculation of expected Auger energies. However, this is no longer necessary since the Auger spectrum is generally used as a fingerprint for identifying elements present in the surface region of the sample under investigation, in much the same way as in XPS. That is, the spectrum from the unknown is compared with standard spectra of the elements, usually from one of the reference handbooks (Davis et al, 1976; McGuire, 1979; Sekine et al,

1982; Shiokawa et al, 1979), and the major peaks identified from their
kinetic energies until all elements in the sample have been located.

It used to be thought that Auger spectral lines were intrinsically broad
and therefore relatively insensitive to the chemical environment of the
emitting atom. However, with modern instruments of higher resolution,
chemical shifts can be observed. There are extensive possibilities for the
determination of bonding information from Auger spectra (Ramaker, 1985) but,
in general, interpretation of the shifts is more complex than in the case of
XPS. This subject is discussed in section 2.4 below.

Angular information from Auger electron spectra can be employed in a
similar manner to that for XPS in order to acquire information on depth
distributions, although, in the case of AES, variation of the spectra as the
angle of incidence is changed is an additional complicating factor in
interpretation of the data. Variations in peak intensities may also occur as
the probing electron beam is traversed from grain to grain across a polycrystalline sample surface. This is due to either channelling of the
incoming beam or diffraction of the emitted electrons along particular
crystal directions. While these effects can be exploited to give structural
information in some cases, in general they represent a source of uncharacterised variation in intensity and can lead to erroneous results, particularly
in the quantification of Auger maps.

In terms of practical surface analysis, AES is preferred over XPS in
situations where high spatial resolution is required. Generally, the samples
need to be conducting and, preferably, tolerant to damage from the incident
electron beam. AES is widely applicable in both fundamental and technological fields, and is perhaps most frequently used in metallurgical and
materials science laboratories and in the microelectronics industry.

2.4 Spectral Interpretation

Interpretation of AES or XPS spectra, and indeed most other forms of
analytical data, will generally proceed from a qualitative to a quantitative
analysis. That is to say, first of all those elements that are present will
be identified and any structural or chemical information noted, and subsequently some estimate of the amounts of each element in the analytical volume
will be made. This second aspect of quantification of AES and XPS spectra is
an extensive subject of much current research activity, some aspects of which
are not yet fully resolved. As such, it is treated separately in Chapter 5.

In AES, particularly where differential spectra are used, the identification of elements is straightforward and simply involves comparing the experimental spectrum with standard spectra from the reference handbooks already
mentioned. Some computer data analysis systems include a library of peak
positions, and can accomplish element identification efficiently and with a
reasonable degree of reliability. For rapidity, the initial identification
should concentrate on the major peaks, following which the minor peaks associated with the major elements present can be identified. Any remaining
unidentified peaks will probably be due to low concentration minor constituents, which can usually be identified by a search of the reference
handbooks.

The qualitative analysis of XPS spectra is more complex than in the AES
case because of the contribution of the background in the undifferentiated
spectra, and the presence of Auger peaks in addition to photoelectron peaks.
This complexity means that in general there may be more information present
than in the Auger electron spectrum from the same area of the same sample,
but the analyst may have to work slightly harder to extract it. As in the

case of AES, the initial problem is to identify the major peaks in the
spectrum and this is accomplished by comparison with reference data (Wagner
et al, 1979). Confusion may occasionally arise where a photoelectron line of
one element occurs close in energy to an Auger line of another. This may be
resolved by taking spectra at two different photon energies, typically
magnesium radiation at 1253.6 eV and aluminium radiation at 1486.6 eV. Auger
lines occur at a fixed kinetic energy, independent of the photon energy,
whereas the photoelectron lines occur at a fixed binding energy with their
kinetic energy dependent upon the radiation used. Therefore, peaks that
shift in apparent binding energy by 233 eV on going from magnesium to
aluminium radiation can be identified unambiguously as resulting from Auger
transitions.

With the major photoelectron and Auger lines in the spectrum identified,
the analyst may be left with a number of minor peaks whose origins are yet to
be ascertained. Such peaks appearing on the high binding energy side of the
principle peaks in the spectrum are almost certainly shake-up lines, multiplet splitting, or due to energy loss processes. Shake-up occurs when photoemission takes place from an ion in an excited state, and can result in a
satellite whose intensity approaches that of the main line. Differences in
the shake-up structure can be used to give information on the chemical state
of the emitting species. The same can be true of multiplet splitting, which
arises when photoemission from a core level of an atom with unpaired electrons in the valence band leaves an unpaired electron in the core level which
can couple with the other unpaired electrons in more than one way. Each
possible configuration has a different energy and results in a different
binding energy component of the photoelectron line. Such effects are most
prominent in the magnetic elements.

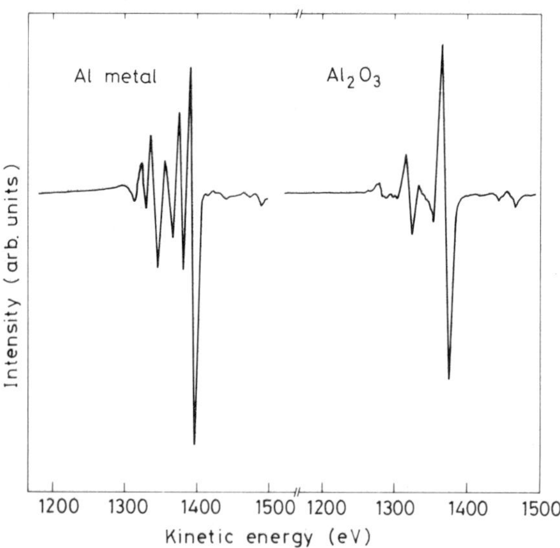

Figure 2.6 Differential Auger spectra from aluminium in the metallic and
oxide state. Large changes in the position and structure of the
loss peaks on the low kinetic energy side of the main peak are
seen, as a consequence of the change in bonding (after Davis et
al, 1976).
(reproduced with permission, Perkin Elmer Inc.)

Energy loss lines occur when the escaping photoelectron or Auger electron interacts with other electrons in the solid, losing a finite amount of energy in the process. This effect can be quite marked in free-electron like metals, where there may be strong coupling to bulk or surface plasmons. A good example is seen in the AES spectra for aluminium in the metallic and oxidised states shown in Figure 2.6. The series of minor peaks on the low kinetic energy side of the main peak are due to plasmon loss features. This is an extreme example, nevertheless, the analyst must be aware of such possibilities giving rise to otherwise unaccounted for spectral structure.

One further source of extraneous peaks in XPS spectra arises from the satellites present in the unmonochromatized radiation from the conventional twin-anode X-ray source described in Chapter 3. The strongest of these can give rise to a peak 8% of the intensity of the main peak at a binding energy lower by 8.4 eV for magnesium radiation or 6% of the intensity shifted by 9.8 eV for aluminium radiation (Wagner et al, 1979). An efficient data system will provide a facility for automatic removal of such satellite peaks.

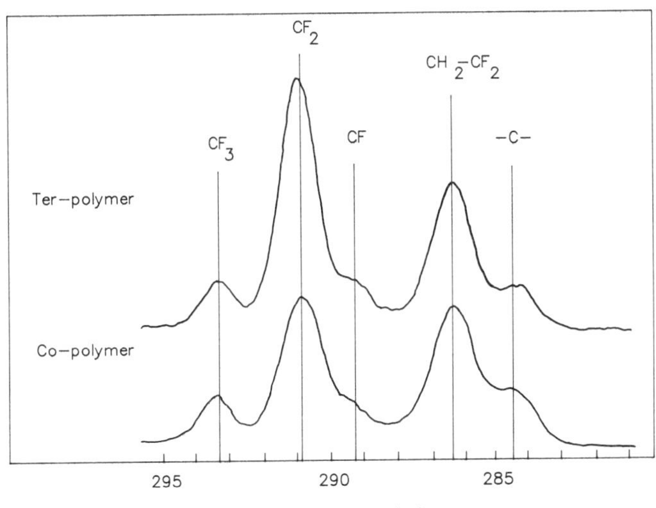

Figure 2.7 The high resolution carbon 1s photoelectron spectrum from the commercially important elastomers hexafluoropropylene vinylidene fluoride and hexafluoropropylene vinylidene fluoride tetrafluoro ethylene. Components corresponding to all carbon bonding configurations present in the elastomer can be distinguished. The C - C bonding present arises from graphitic filler (after Martensson, 1988).

An important aspect of XPS, to a greater extent than AES, is the identification of the chemical state of the emitting species. Essentially, the stronger the chemical bonding experienced by an ion, the more tightly bound its own energy levels become and, correspondingly, the higher the binding energy (or the lower the kinetic energy) of the photoelectron or Auger peaks from those levels. In general, the more tightly-bound the atom is in a particular chemical configuration, the higher will be the binding energy of the corresponding peak in the photoelectron spectrum. The change in energy, ΔE, of the photoelectron line from atom A, known as its chemical shift, is related to the local electronic environment by:

$$\Delta E = k_A q_A + \sum_B {}'(q_B/r_{AB}) + L \qquad [2.2]$$

where q_X is the charge on atom X and r_{AB} is the sphere radius for atom A and its neighbour atom B. L and k are constants determined by experiment (Gelius, 1974). The second term in equation [2.2] is the Madelung potential at site A, for a solid material.

As an example, Figure 2.7 shows the XPS spectrum of carbon in the copolymer hexafluoropropylene vinylidene fluoride and the terpolymer hexafluoropropylene vinylidene fluoride tetrafluoroethylene, used in the production of fluorocarbon elastomer materials (Martensson, 1988). Distinct peaks at clearly separable energies are seen, corresponding to emission from carbon atoms in each of the possible bonding states of the molecules. A further component at approximately 285 eV originates from the carbon of the graphite filler used in the commercial form of the material. Outside the polymer industry, commonly observed shifts include those due to various oxidation states of the metallic elements. Figure 2.8 shows an XPS spectrum of the silicon 2p emission line acquired from a silicon (100) single crystal slice. The two peaks are due to pure silicon and to the presence of a thin adventitious overlayer of silicon in the oxide state. The silicon $2p_{1/2}$ and $2p_{3/2}$ peaks are not resolved in this example. A shift of approximately 3.2 eV to higher binding energy is observed for the component arising from the oxide relative to the pure element.

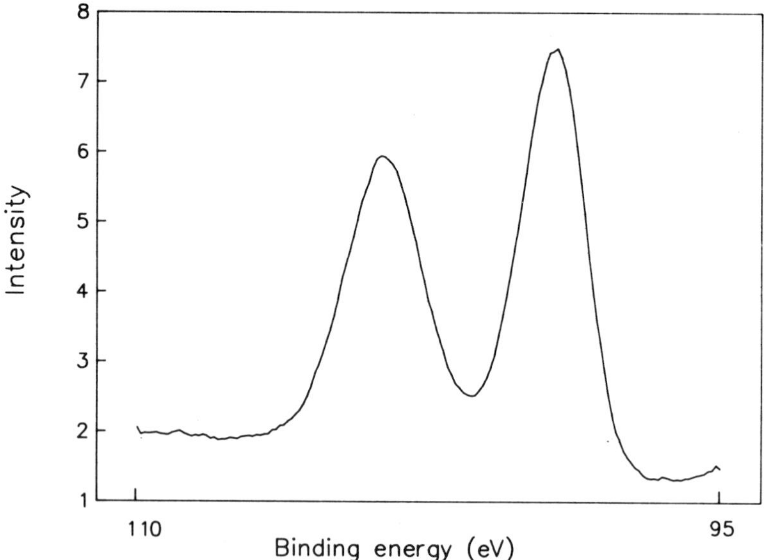

Figure 2.8 XPS spectrum of the the surface of a silicon (100) crystal slice showing splitting of the silicon 2p line into components due to the pure element, at a binding energy of approximately 99 eV, and the adventitious oxide layer at approximately 102 eV.

There is a voluminous literature concerning the interpretation of
chemical states from XPS spectra. For a review see e.g. Dilks (1981).
Unfortunately, some of the early work in this field was performed using
inadequately calibrated spectrometers and is consequently unreliable. To
assist in the use of the literature, the National Institute of Standards and
Technology (NIST, Gaithersburg, MD, USA) has produced a database containing
around 13000 chemical state records taken from the literature (National
Institute of Standards and Technology, 1989). These are given a classific-
ation showing their degree of reliability, and may be used with confidence.
The NIST database includes records up to 1985, however, much work of high
quality has been performed since that date and updates are expected

Table 2.1 Binding energies for C 1s and O 1s photoelectron
lines in the presence of carbon and oxygen bonding, after Desimoni
et al (1990).

Carbon atom chemical environment	C 1s binding energy (eV)
graphite, aromatic hydrocarbon	284.6
aliphatic hydrocarbon	285.1 - 285.3
alcohols, phenols	286.1
carbonyl	287.6
carboxylic acids	289.1
carbonate, CO_2	290.6
plasmon	291.3

Oxygen atom chemical environment	O 1s binding energy (eV)
alcohol C-OH in carboxylic acids	532.3 - 533.3
ketone C=O in carboxylic acids carbonate	531.1 - 531.8
water	535.5 - 536.1

In addition to the chemical shifts of XPS lines due to atoms in a
particular chemical environment or bonding configuration, secondary shifts
can be observed for atoms adjacent to those showing the primary shift. Such
effects may be observed in the C 1s spectra of oxygen-containing polymers
where, for example, the component due to carbon in a carbonyl bond will show
a relatively large and easily distinguishable peak but high resolution
studies show the presence of a less-strongly shifted component due to the
nearest neighbour carbon atoms. In a systematic study of secondary shifts in
spectra from oxygen-containing polymers using instrumentation of the highest
available energy resolution, Briggs and Beamson (1992) found secondary shifts
of typically around 0.4 eV, extending up to 0.7 - 0.8 eV in polymers

containing methacrylate groups. Their work, covering a very wide range of
organic polymers, has also been published in book form (Beamson and Briggs,
1992).

As a final example of chemical state identification, Desimoni et al
(1990) have taken the additional step of correlating the structure within
individual peaks with the overall elemental composition of the sample. In
their work on carbon fibre surfaces, they force the fraction of intensity in
the carbon 1s peak attributed to oxygen bonding to equate to the same overall
level of oxygen in the sample as determined from the oxygen 1s peak.
Similarly, they require the fraction of intensity in the oxygen 1s peak
attributed to carbon bonding to give the same carbon content as that derived
from the carbon 1s peak. While relatively straightforward in the case of a
sample with only two elements of interest, the application of this technique
to multi-element specimens would pose a serious challenge to the analyst.
Nevertheless, such an approach imposes a degree of rigour that greatly adds
to the credibility of the analytical result. This work also shows an example
of the use of the carbon KVV Auger line shape and position to obtain information on the relative probability of the presence of sp^2 or sp^3 bonding. The
peak positions derived by Desimoni et al (1990) for the carbon and oxygen 1s
photoelectron lines in the presence of different degrees of carbon to oxygen
bonding are shown in Table 2.1.

Because of the difficulties of accurate photoelectron and Auger line
position identification in XPS, Wagner (1975) has suggested the use of a
modified Auger parameter as an aid to chemical state identification. The
modified Auger parameter, α', is defined as the sum of the kinetic energy of
the principal Auger peak and the binding energy of the principal photoelectron peak in the XPS spectrum of interest. This is a useful quantity as
it varies with chemical state but does not depend on an absolute measurement
of energy, and is therefore independent of the charge state of the sample. A
data set is available which provides values of α' for a wide range of
elements and compounds. Use of this should enable the analyst to make a
reasonably unambiguous assignment of chemical state in a large number of
cases (Wagner, 1990).

In addition to element and chemical state identification, the spectra
also contain data on the depth distribution of the species contributing to
the emission. Interpretation of these data to give non-destructive depth
profiles is an important and growing area of application for both XPS and AES
and, as such, is discussed separately in Chapter 6.

CHAPTER 3

INSTRUMENTAL TECHNIQUES FOR XPS AND AES

3.1 General Requirements

AES and XPS involve respectively the interaction of an electron or X-ray photon beam with the specimen surface to be analysed, followed by detection of Auger elctrons or photoelectrons. If the incident beam is to reach the sample, and the emitted electrons are to be detected, then their mean free paths in the region of the sample must be greater than the physical dimensions of the apparatus involved, otherwise scattering will distort the outcome of the experiment. For physically realistic dimensions, this implies the use of a vacuum. Basic considerations of the kinetic theory of gases lead to the conclusion that, for apparatus with dimensions of the order of a few tens of centimetres, pressures in the 10^{-5} to 10^{-6} mbar range would be adequate. However, for surface science there is a further and much more stringent requirement to be satisfied, namely that the surface to be examined does not become contaminated by the residual gas in the vacuum chamber during the course of the analysis. This applies even to surfaces which have already been air exposed, as the analysis will often include depth profiling in which the contaminated outer surface layers are removed revealing a fresh surface to be analysed.

In the worst case, every gas particle that arrives at a clean surface will adhere and contribute to surface contamination. For typical surface densities and gas atom types, a full monolayer of contaminant atoms will form in around one second at a pressure of around 10^{-6} mbar. This is known as a gas dose of one Langmuir. The unity sticking coefficient assumed here for the gases typically present in the residual vacuum of a surface analysis system will be valid only for the more highly reactive surfaces, however, this does show the need for pressures in the 10^{-10} mbar range if contaminants are to be kept below a few percent during the course of a typical experiment. In practice, unless very careful argon-ion depth profiling is involved, routine surface analyses can be performed satisfactorily in vacuum systems with base pressures in the low to mid 10^{-9} range.

Pressures of this order are known as ultra-high vacuum (UHV). The advent in the late 1960s of a reliable UHV technology, based on bakeable stainless steel chambers with access ports sealed by copper gaskets, proved a great spur to the development of surface science, which had previously relied on all-glass systems with their associated inflexibility and time-consuming construction. The achievement of UHV is routine in all surface analytical

laboratories. Good texts on the subject exist (e.g. Redhead and Kornelsen,
1993), to which the interested reader is referred for more details. It is
worthwhile to note that samples are commonly introduced into the UHV environment through air-lock systems which allow the transition from atmospheric
pressure to UHV to occur in a few minutes.

It should be pointed out that the need for UHV does restrict the type of
surfaces that can be investigated using these techniques. For example, many
biological samples, the structures of interfaces under liquids, or the
surface composition of high vapour pressure non-UHV compatible solids cannot
be studied. The search for methods of surface analysis which do not need
vacua will be a growth area in the future.

For AES and XPS, the minimum apparatus required is a UHV chamber
equipped with some kind of sample mount, a means of producing the incident
beam and directing it onto the sample surface, and a means of analysing and
detecting the emitted electrons. Often, the two techniques are found in a
combined electron spectroscopy system equipped with an electron gun, an ion
gun for depth profiling, and an X-ray source, together with an electron
energy analyzer of the electrostatic deflection type. Such a system may
typically also contain a precision x-y-z sample manipulator probably with
sample rotation about its central axis, a scintillator for detection of
secondary electrons enabling a scanning electron microscope (SEM) image of
the sample to be formed, possibly an X-ray detector for simultaneous bulk
analysis by electron-probe microanalysis, together with viewports, pumping
lines, vacuum gauges and a means of providing rapid sample entry.

Such systems represent a large investment, both in capital equipment and
in the highly trained staff required for their operation. Their widespread
use in all the developed countries of the world testifies to the demand for
surface analysis and to the benefits obtained from its application to
technological and industrially significant problems.

In this chapter the methods of producing the necessary electron, photon
and ion beams are discussed in section 3.2. Energy analysis of the emitted
electrons is described in section 3.3. The effects of radiation damage and
sample charging are covered in sections 3.4 and 3.5, and the achievable
spatial resolutions of the techniques are discussed in section 3.6.

3.2 Excitation Sources for Surface Analysis by XPS and AES

3.2.1 X-ray Sources

X-ray sources for XPS are generally based on the design of Henke (1963),
in which thermionic electrons emitted from a tungsten filament are accelerated and focussed onto a water-cooled anode. The impact of the high energy
electrons causes core-level holes to be created within the electronic structure of the atoms of the anode material. These then relax by the emission of
X-rays of characteristic energy. For efficient cooling, the anode is constructed from copper. Line of sight between the filament and the emitting anode
surface must be prevented, otherwise contamination by evaporated tungsten
will occur, and the sample must similarly be protected from contamination by
the presence of a differentially-pumped aluminium window sufficiently thin
(typically 3 μm or less) to transmit the X-rays produced.

The copper characteristic X-rays are not of practical use in surface
analysis, therefore alternative materials such as magnesium or aluminium are
deposited onto the anode, giving K series X-rays at principal energies of
1253.6 eV and 1486.6 eV respectively. The twin-anode design shown in Figure

3.1 is popular and enables either Mg or Al Kα X-rays to be produced by energizing the appropriate filament. The availability of two different photon energies is useful since it enables a distinction to be made between Auger photoelectron lines in the spectra, as explained in Chapter 2.

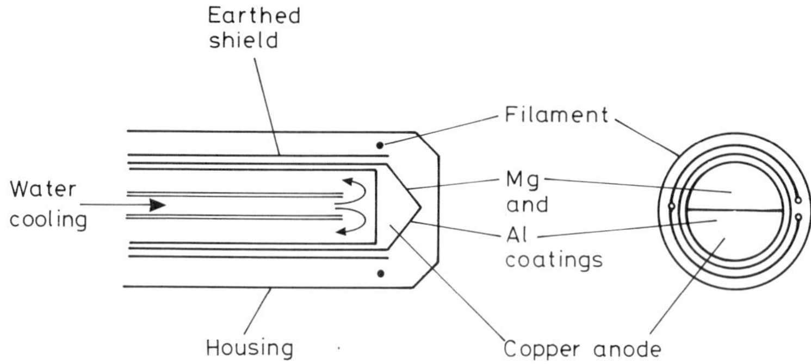

Figure 3.1 A schematic cross-section of a typical twin anode magnesium and aluminium X-ray source. Energising either the upper or lower filament results in thermionic electron emission and subsequent acceleration onto the magnesium- or aluminium-coated face of the anode, respectively.

It is difficult to quantify the output of such devices but typically a photon flux of the order 10^{12} per second into an area of around 1 cm^2 on the sample may be produced. A power supply giving up to 15 kV positive potential on the water-cooled anode (a difficult design problem when leakage currents and operator safety are considered), together with an emission current stabilized filament supply of around 5 - 6 A capacity is required. Variations on the basic design include alternative anode materials such as silicon or silver, the use of several rather than two anodes, and high speed rotating anodes for greater power dissipation and hence higher photon fluxes on the sample.

Advantages can result from the use of a monochromator in conjunction with the X-ray source. Typically, a curved crystal is used in a Rowland circle geometry with either 0.5 or 1 m radius. This gives focussing and monochromatization of the Al Kα line, although with a significant loss in overall flux on the sample. This need not be a disadvantage if the electron optics of the analyzer are correctly optimised as described in section 3.5 below. The use of a monochromator is particularly important where high energy resolution is required in the study of subtle lineshape changes or chemical shifts. Resolutions as high as 0.3 eV have been reported (Perkin Elmer, 1990) for silicon 2p emission, comparing very favourably to the natural unmonochromated line width of the Al Kα line of approximately 1 eV.

As an alternative to laboratory sources, synchrotron radiation is occasionally used. The variable photon energy available can be used to enhance emission from different atom types present in the sample by making use of their different cross-sections for photoemission. Alternatively, it can be exploited to vary the probe depth of the experiment through the variation in escape depth of photoemission lines at different kinetic energies. However, at present synchrotron radiation is not available from laboratory sources (although attempts to build a "bench top" synchrotron have been made), and expensive central facilities must be used. For this reason

the vast majority of XPS spectra are taken using the twin-anode magnesium-
aluminium type of source described here. The use of synchrotron radiation is
restricted to situations in which its properties of tunability and high
brightness are essential to the success of the experiment.

3.2.2 Electron Beam Sources

For AES, stable electron beams of energies 2.5 kV to 30 kV focussed into
a small spot and addressed to various points or raster-scanned across the
sample surface are required. Originally, the electron guns used in surface
science were simple adaptations of the guns used in TV tubes. However, the
inevitable push towards higher spatial resolution and brightness has lead to
a number of technical innovations in the electron optical design and in the
type of emitter used.

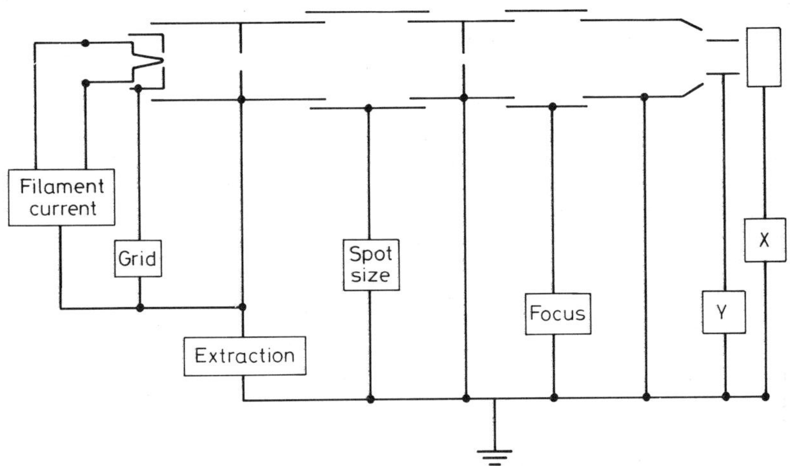

Figure 3.2 Schematic layout of a modern electrostatically focussed electron
gun. The electrodes have circular symmetry (except for the scan
plates), and the gun may be designed to fit through a standard 70
mm UHV chamber port.

Figure 3.2 shows an outline of the electron-optical scheme of a typical
modern electrostatically focussed electron gun. Electrons are thermionically
emitted from a directly heated tungsten hairpin filament, with the emission
stabilised by feedback from the grid or Wehnhelt electrode. Electrons are
extracted from the small source formed in this region by the high positive
field from the first anode, beyond which they are focussed and projected onto
the sample. Two stages of focussing are used to give control over the spot
size and beam divergence separately, with the spot positioning and scanning
controlled by deflector plates usually mounted just outside the final lens
element. All the electronic supplies are connected to scale with the beam
voltage (the high tension applied to the filament) so that the focus is
retained as the beam energy is adjusted. In addition to that shown, one or
more sets of deflection poles may be incorporated to correct for astigmatism
in the beam. A well designed gun of this type may typically operate with
beam voltages up to 10 kV and give a spot size of 200 nm. The beam current
at the highest resolution may be in the nanoamp range, increasing up to
perhaps a few microamps if the spot size is greatly degraded.

Further improvements possible to such guns include the use of single crystal Lanthanum hexaboride as the emitting surface. The lower work function, compared to tungsten, and the precisely controlled geometry, give a much higher brightness when used as a direct replacement for a tungsten hairpin. Even better performance may be obtained by the use of a field emission source, where an etched tungsten tip is used as the emitting surface. Such sources must always be kept in UHV and require some form of differential pumping. However, very high brightnesses, in the range of 1 nA into a 20 nm spot at 30kV, can be obtained. The various possible combinations of spot size and beam current available from different design approaches have been reviewed by Seah (1986).

For beam energies beyond the 10 kV range very high focussing and deflection voltages are required, and electrostatic guns become increasingly difficult to design and use. For energies in the range 20 - 30 kV, or occasionally up to 100 kV, magnetic systems are preferred. At this point, surface science and electron microscopy technologies converge with the addition of electron energy analysis to UHV scanning and scanning-transmission electron microscopes (STEM) (Venables et al, 1986). Spot sizes of 1 nm may be achieved, and surface analysis may be only one of a range of facilities offered by such an instrument, whose construction is dominated by the electron gun column. One disadvantage of magnetic focussing is the presence of residual magnetic fields in the sample region. These can distort the low energy end of the electron spectrum, giving potentially misleading results unless the effect is understood and compensated for. Recent designs for very high spatial and energy resolution systems include the use of magnetic field parallelizers to extract electrons to an electrostatic analyzer remote from the sample region.

3.3 Electron Energy Analyzers for AES and XPS

In order to perform any electron spectroscopy, it is necessary to energy analyze and subsequently detect the electrons emitted from the sample surface. To do this, the electrons must be passed through an electron optical component that either acts as a filter or exhibits the property of dispersion. Early methods of Auger electron detection relied on the use of a low energy electron diffraction (LEED) optics re-configured as a high pass energy filter. For a description of the LEED technique see Clarke (1985). Modern instrumentation dedicated to XPS or AES almost exclusively uses electrostatic deflection analyzers which spatially disperse electrons across some detection device according to their kinetic energy.

The nature of the sources used dictates that different characteristics are required of analyzers for AES and XPS. In AES, electrons must be collected from a single point of emission on the sample surface, at the position of the probing electron beam, and analyzed with a resolution of typically 0.3% of the kinetic energy. For maximun signal, as large a solid angle of collection as possible is required. In conventional XPS, maximum useful signal intensity is obtained when electrons are collected from the whole of the broad emitting area and analyzed with a constant resolution generally in the range 1 - 2 eV. Often it is desirable to restict the solid angle of collection in order that angle dependent studies may be carried out. Although AES and XPS instrumentation evolved through different routes, it is common practice today for both facilities to be built into one instrument, using the same energy analyzer operating in different modes for each spectroscopy.

The detection of Auger electrons was first demonstrated by Langer (1953) with the use of a 90° spherical sector electrostatic analyzer and subsequently by Thorp and Scheiber (1967) using a retarding field analyzer

(RFA). The original XPS instrumentation of Siegbahn made use of magnetic sector analyzers, adapted from mass spectrometer designs. However, present day analyses by XPS and AES are almost universally carried out using a version of the cylindrical mirror analyzer (CMA) or hemispherical sector analyzer (HSA). An excellent introduction to the characteristics and general properties of electron spectrometers is given by Roy and Carette (1977).

Electrostatic energy analyzers fall into two generic types, the "mirror" and the "deflector", exemplified by the CMA and the HSA respectively. It has been pointed out that the CMA and HSA are both extreme types of a general class of toroidal electrostatic analyzers (Leckey, 1987), nevertheless it is convenient to treat them separately here. Originally, the CMA was evolved primarily for use in AES whereas HSA designers were principally motivated by performance in XPS. These distinctions are now blurred, with modified versions of each type of analyser being used for either, or both, techniques.

3.3.1 Cylindrical Mirror Analyzers

An outline of a simple CMA design is shown in Figure 3.3. The operation of the analyzer is described by Sar-el (1967) and reviewed at various levels in a number of texts (Seah, 1988; Riviere, 1990; Woodruff and Delchar, 1986). The basic construction consists of a pair of coaxial cylinders with entrance and exit slits, and a detector. An electric field between the two cylinders deflects electrons onto the detector by an amount dependent on their kinetic energy. The detector is usually a single channel electron multiplier, although other types of electron multiplier or position sensitive detector have been used, Scanning the electric field sweeps electrons of different energies across the detector, thus generating the spectrum.

The instrument is normally used in an optimised condition in which the source of electrons to be analyzed is aligned with the common axis of the coaxial cylinders, and positioned so that the mean entrance angle is 42.3°. All important parameters of the analyser scale with the inner radius r_1 and a second order focus is formed at a distance L from the sample given by

$$L = 6.13r_1 \qquad\qquad\qquad [3.1]$$

The maximum axial distance reached by electrons that pass through the output slit is given by

$$r_m = 1.80r_1 \qquad\qquad\qquad [3.2]$$

and, obviously, r_2 must be greater than this.

The analysis is performed by sweeping the negative voltage applied to the outer cylindrical mirror, while holding the inner cylinder at earth potential and, with the optimized conditions described, the relationship between the applied voltage V and the energy of the electrons analyzed is given by

$$E = 1.31 \: / \: \ln\,(r_2/r_1) \quad eV \qquad\qquad\qquad [3.3]$$

An important parameter governing the performance of any energy analyser is the resolution, generally taken as the full width at half the maximum intensity (FWHM) of the output signal that would be obtained for a perfectly monochromatic input. For the CMA this is approximately given by

$$(\Delta E/E) = (0.18w/r_1) + 1.39\,(\alpha)^3 \qquad\qquad\qquad [3.4]$$

where α is the semi angular acceptance, typically 6°, and w is the width of
the output slit. In practice the effects of finite source size, aberrations,
stray electric and magnetic fields, and non-uniformity of the meshes used to
cover the apertures, combine to degrade the measured resolution by around a
factor of two from that expected from the above equation. A good CMA
designed along these principles may be fitted with interchangeable slits to
give resolutions of probably 0.3% and 0.6%.

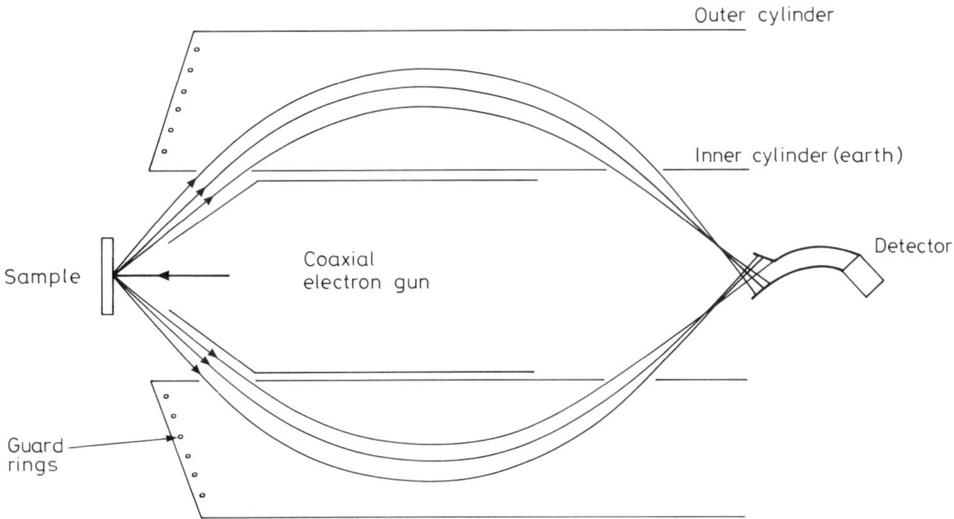

Figure 3.3 Schematic cross-section of a simple single-pass CMA fitted with an
internal co-axial electron gun for AES

The FWHM of peaks in the energy spectrum measured by the CMA increase
with energy approximately as given above, however, the angular acceptance of
electrons emitted from the sample is defined by the input slit geometry and
remains constant with energy. Consequently, the transmission function of an
ideal CMA is proportional to the kinetic energy, E. That is to say, the
measured spectrum is not simply n(E) but is En(E). There is an additional
factor to be considered in a discussion of the response of the instrument,
namely the energy dependent detection efficiency D(E) of the electron
multiplier or other detector used. Thus, the output current I(E) measured as
the outer mirror voltage is varied is given by

$$I(E) \propto En(E)D(E) \hfill [3.5]$$

and this may be further complicated by the presence of stray fields,
incorrect sample positioning and the transfer function of the electronic
measuring system employed (Seah and Smith, 1990). Theoretical detector
efficiencies for the usual types of single channel electron multiplier
("channeltron") found in these systems are given by Seah (1990), and
experimental curves derived from an interlaboratory comparison of Auger
electron spectroscopy instruments involving over 40 laboratories worldwide
have verified the accuracy of the calculations (Seah and Smith, 1991). It
has been demonstrated that the D(E) correction term can be much reduced by
the use of a pair of high transparency grids with an applied bias voltage
between the output slit and the electron multiplier (Seah and Smith, 1990).

In practice, the CMA is used in one of three modes to generate either direct spectra by pulse-counting or analogue multiplication, or differentiated spectra. With a low primary beam current in the nanoamp range insufficient signal is produced for direct current measurement, and the detector is operated in the pulse counting mode. With a high voltage applied to the channel electron multiplier, each incident electron generates a single output voltage pulse which may be amplified, shaped and counted to give a measure of the signal at that particular energy in the spectrum. At higher beam currents, the electrons arrive at the multiplier too closely spaced to be distinguishable as separate events and, with a lower applied voltage, a d.c. output current is generated. Sometimes, in this analogue mode, the incident electron beam is chopped and a lock-in amplifier, tuned to the chopping frequency, is used to detect the output current of the multiplier. This so-called beam brightness modulation or beam blanking mode gives the direct spectrum with a good signal to noise ratio. To obtain the differentiated spectrum a small modulation voltage is added to the outer mirror voltage and the lock-in amplifier is used to detect the component of the output signal in phase with the applied modulation. It is important to note that in going from the pulse counting mode to the analogue multiplication mode the energy dependence of the detector efficiency will change and hence the apparent shape of the spectrum with it.

A version of the CMA, first described by Palmberg (1975), has been produced which uses two sets of cylindrical mirrors coupled with pre-retardation by a grid system. This combines high transmission with the ability to function in either constant $\Delta E/E$ or constant ΔE modes, whereas the traditional CMA is restricted to constant $\Delta E/E$ operation. In general, this double-pass CMA design will give better performance than a concentric hemispherical analyzer and lens combination, of the type discussed below, for medium and poor resolution conditions, but suffers from a loss of intensity at higher resolutions. The problems of specimen alignment and access of probe beams to the sample, inherent in the conventional CMA design, also remain in the double-pass instrument.

3.3.2 Hemispherical Sector Analyzers

Although the CMA is a very successful instrument design for use in AES it is not ideal for XPS. The intrinsic resolution is not especially high, and degrades if a broad source of electrons is to be analyzed, as is usually the case in XPS. For XPS, better performance is obtained if an analyzer of the concentric hemispherical sector type is used. A schematic outline of a 180° spherical sector analyzer is shown in figure 3.4. As was the case for the CMA, the behaviour of such a system has been extensively reviewed by a range of authors (Roy and Carette, 1977; Seah, 1988; Riviere, 1990; Woodruff and Delchar, 1986).

The electron optical focussing properties of a spherical condenser were first described by Purcell (1938), whose equations were subsequently used by Kuyatt and Simpson (1967) to build quite a sophisticated electron monochromator/analyzer combination with an energy resolution of the order of 0.02eV.

As shown in Figure 3.4, the hemispherical analyzer in its simplest form consists of conducting inner and outer hemispheres of radii r_1 and r_2 respectively, with input and exit slits of widths w_1 and w_2. Electrons entering on-axis at the input slit are deflected by an electric field between the two hemispheres in such a way that only those electrons travelling at the pass energy $E_p = eV_p$ of the analyser follow a path given by the mean radius r_o and arrive at the output slit. Thus the device exhibits the property of dispersion and can be used as an energy analyzer. The voltage required to analyze electrons at the pass energy is given by

$$V = V_p \left[(r_2/r_1) - (r_1/r_2) \right] \qquad\qquad [3.6]$$

and the potentials of the inner and outer hemispheres are $V_p[3-2r_o/r_1]$ and $V_p[3-2r_o/r_2]$ respectively. The energy resolution (FWHM) of such a device is given by

$$(\Delta E/E)) = \left[(w_1 + w_2)/2r_o \right] + \alpha^2 \qquad\qquad [3.7]$$

where α is the maximum angle electrons make with the axis. Electrons are preretarded to the pass energy E_p before analysis, therefore for a given pass energy the energy resolution ΔE is constant.

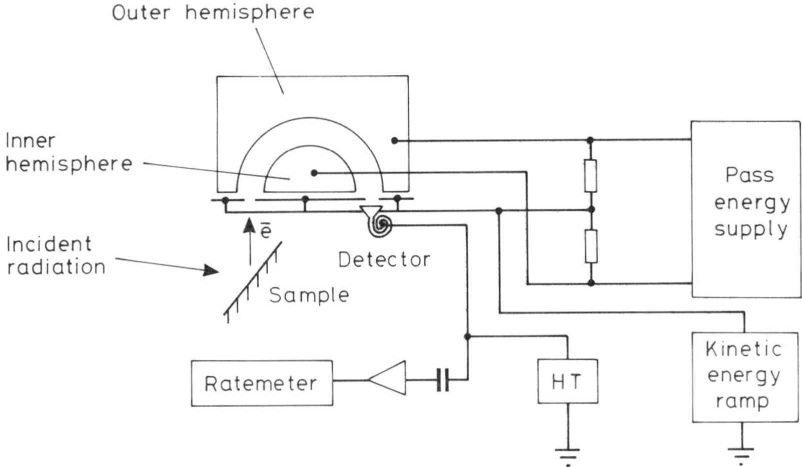

Figure 3.4 Schematic arrangement of a simple hemispherical sector electron energy analyser. Also shown is one possible arrangement of the electronic system necessary to power such a device.

Hemispherical sector analyzers are rather robust, electron optically. In practice, sectors of perhaps 150° may be used, while maintaining point-to-point focussing, with no loss in performance. Further, it is often found convenient to truncate the out-of-plane sector angle for engineering reasons, again with no apparent degradation in performance.

In modern instrumentation for XPS and AES, the hemispherical sector analyzer is almost always used in conjunction with an electron optical input lens. This gives the simple practical advantage that the analyzer is then remote from the sample, allowing better sample access for the various probe beams and manipulators required. From the designer's point of view it also enables the analyzed area on the sample to be optimised to match the source, and allows either constant fractional resolution ($\Delta E/E$) or constant energy resolution (ΔE) modes to be employed. The basic hemispherical sector analyzer with an input lens is shown in figure 3.5. The lens illustrated is of the simple 3-element Einzel type whose properties are well known; in commercially-available instruments a variety of input lens types are used but the principle rationale for their use remains the same.

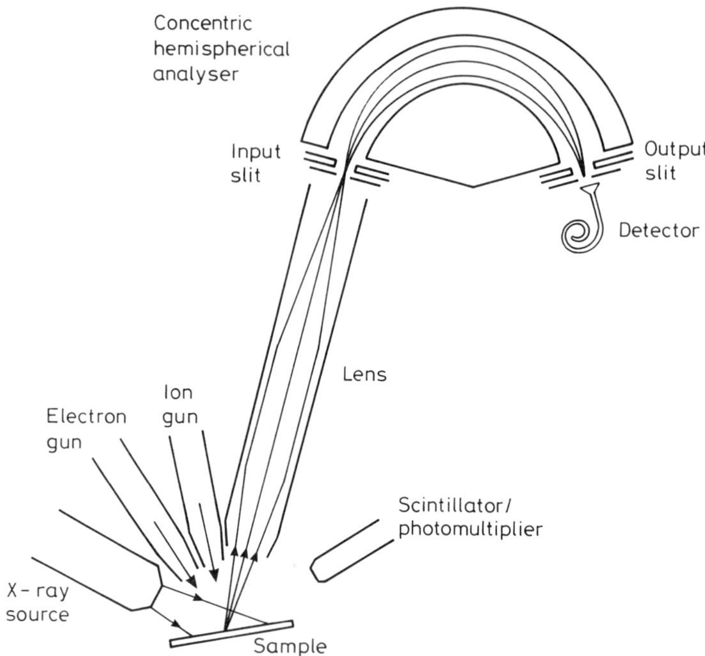

Figure 3.5 A typical arrangement of a hemispherical sector electron energy analyser fitted with an electron-optical input lens, together with a variety of excitation sources such as may be found in a multi-technique surface analysis instrument.

If it is desired to operate the lens/analyzer system in such a manner as to generate spectra with constant energy resolution ΔE, then the pass energy of the analyzer is held constant and the lens is used to focus electrons of the required energy from the sample to the input slit of the analyzer. This is the normal mode of operation in XPS, and is known as the constant pass energy, constant analyzer energy, constant ΔE, or the fixed analyzer transmission mode. If constant fractional resolution $\Delta E/E$ is required, then the lens can simply be used to transport electrons to the input of the analyzer whereupon they are analyzed in the conventional way to give a resolution described by equation [3.7] above.

Typical values might be a mean radius of 100 mm with input and output slits of 2 mm width and an angular acceptance of 5°, in which case evaluation of equation [3.7] gives a resolution of 2.8% of the pass energy, i.e. 2.8 eV for 100 eV electrons and 28 eV for 1000 eV electrons. However, if the input lens is used to retard the electrons before analysis then the figures are much improved. For example, retardation by a factor of 10 will give a theoretical resolution of 0.28 eV for the 100 eV electron and 2.8 eV for electrons at 1000 eV. This method of operation is referred to as constant $\Delta E/E$, constant retard ratio, or fixed retard ratio. Alternatively, all the electrons could be retarded to a pass energy of 50 eV, giving a constant energy resolution of, in principle, 1.4 eV. Care must be taken not to apply excessive retardation, as analysis at too low a pass energy can result in spectral distortion due to internal scattering of secondary electrons in the analyser (Seah and Smith, 1990).

In the constant $\Delta E/E$ mode, the transmission of the analyzer is
proportional to the kinetic energy, E, making it the favoured mode for AES
where the high energy peaks tend to be weaker. In the constant ΔE mode the
transmission is a complex function of the geometry and precise operating
conditions of the analyzer but is generally approximately constant for low
energies and falls off, initially as $E^{-0.5}$ and subsequently as E^{-1} at higher
energies (Seah, 1980), with smooth transitions between the three regimes.

The usual detector used in a CHA-based system is a single channel
electron multiplier, similar to that used in the CMA, with its associated
energy-dependent detection efficiency. In the constant $\Delta E/E$ mode of
operation the signal intensity is given by the analogous equation to [3.5]:

$$I(E) \propto En(E)\ D(E/R) \qquad [3.8]$$

Here, the kinetic energy of the electrons has been reduced by the lens to a
fraction 1/R of the initial value. Analysis takes place at energy E/R and
consequently electrons strike the detector at that energy. However, when the
analysis is carried out in the constant ΔE mode, the electrons all strike the
detector at the same energy, the pass energy of the analyzer, and the energy
dependence of the detection efficiency does not affect the shape of the
spectrum. Unfortunately, the transmission of the analyzer is not a simple
function of energy. Therefore, for the constant ΔE mode:

$$I(E) \propto n(E)\ T(E) \qquad [3.9]$$

where T(E) is the energy dependent transmission function.

A further advantage of the CHA and lens combination is that careful
choice of the lens magnification and input slit size gives control over the
area on the sample from which electrons are accepted. This is exploited in
small-area XPS (see section 3.5 below).

3.3.3 Multichannel Detection

Where a single channel electron multiplier detector is used in a HSA or
other disperive analyzer, an output slit is required in order that energy
resolution may be achieved. This results in a loss of potential signal,
depending upon the ratio of the slit width to the gap between the
hemispheres. A number of ingeneous methods have evolved to utilize this
"lost" signal. These generally involve partially or completely removing the
output slit and using some kind of position sensitive detector system. The
position of arrival of an electron on the output plane of the analyzer is
dependent on its kinetic energy, therefore, if positional information can be
retained across a detector in the ouput plane, a small section of the
spectrum may be collected in parallel, greatly increasing the output signal.

The simplest multichannel detection system is a number of single channel
electron multipliers placed side by side across the output slit, each with
its own preamplifier and counting circuit. For a 3-channeltron system the
central channeltron detects electrons at the pass energy E_p while the outer
channeltrons collect at $E_p - \delta E$ and $E_p + \delta E$. During scanning the computer
data acquisition system accumulates counts into appropriate channels as the
kinetic energy is varied through the spectrum and, at least in theory, a
factor of three improvement in count rate is obtained.

A number of alternative approaches to multi-channel detection in
electron spectroscopy have been explored. Most of these use micro-channel
plates as pulse-amplifiers across the whole of the output slit width,
followed by some position sensitive detector which records the position and
intensity of the micro-channel plate output. This could be a phosphor

screen, converting the electron intensity distribution to light, followed by
a TV camera (Schnell et al, 1984; Morris et al, 1984). Alternatively, a
light sensitive semiconductor device such as a charge-coupled device (CCD)
(Hicks et al, 1980) or photodiode array can be used to convert the image on
the phosphor to an electronic signal. A method which does not use the
phosphor screen intermediary is the resistive anode strip (Pollard et al,
1982). The resistive anode is useful in cases where signal intensities are
low, for example in high energy or angular resolution studies, but cannot be
operated at the higher count rates possible in more usual XPS and AES
experiments.

Commercially-available variants of some of these position-sensitive
detection systems have been made available. However, at present it appears
that the simple multi-channeltron system, probably with three or five single
channel electron multipliers, is actually the most effective in routine
analytical use despite lacking the conceptual elegance of some of the
alternative methods.

3.3.4 Analyzer Calibration

For modern analytical electron spectroscopy, it is essential that the
spectrometer is properly calibrated, otherwise reliable and reproducible
results cannot be assured. If surface analysis is to take place in an
accredited laboratory, then calibration must be built in as part of the
overall quality assurance system. In XPS and AES, calibration concerns both
the energy and intensity scales of the spectrometer. Furthermore, it is
necessary to verify the linearity of the intensity scale. The energy scale
may be calibrated using reference line energies traceable to atomic
standards, whereas calibration of the energy dependence of the intensity
scale (the transmission function of the instrument) is, in principle, made by
comparison with published standard reference spectra.

The problem of energy calibration was first highlighted by results from
interlaboratory comparisons organised by the American Society for Testing
Materials (ASTM) (Powell et al, 1979; 1982). Binding energies in a range of
2 eV were reported for the Cu $2p_{3/2}$ photoelectron peak at around 933 eV,
although typical repoducibilities within one instrument may be two orders of
magnitude better than this. For AES, a significantly worse scatter of 16 eV
was found for the Cu L_3VV line. Precision of this poor level is totally
inadequate if chemical state data measured in one laboratory and reported in
the literature is to be of use to workers in other laboratories using other
instruments.

To rectify these problems, the UK National Physical Laboratory (NPL)
embarked upon a program of research to define reference values for XPS and
AES peak energies, traceable to internationally agreed primary standards.
Using a precision potentiometer to compare the spectrometer retarding voltage
with standard reference cells, it was possible to provide absolute values for
calibration peak energies with accuracies in the range 0.01 to 0.02 eV for
XPS (Anthony and Seah, 1984). For AES the absolute accuracy acheived was
slightly worse, at 0.02 to 0.04 eV (Seah et al, 1990). The utility of these
results was subsequently verified by interlaboratory comparisons (Anthony and
Seah, 1984). Spectrometers calibrated for XPS using the Au $4f_{7/2}$ and Cu
$2p_{3/2}$ peaks were found to have a typical mid-range error of only 0.035 eV
(Seah, 1985). In AES the situation is more complex as spectra may be
recorded in either direct or derivative mode, with a wide range of energy
resolutions. Nevertheless, using the Cu $M_{2,3}$VV and L_3VV peaks it was found
possible to maintain an excellent calibration (Seah and Smith, 1990). Table
3.1 shows calibration values from the NPL work for XPS peak energies, using

magnesium and aluminium K radiation, corrected to the post-1990 representation of the SI volt (Seah, 1989).

The accurate placement of peaks on the spectrometer energy scale allows element and chemical state identification to proceed with confidence. However, if quantitative information is required, it is essential to be confident that the spectrometer output is an accurate representation of the spectral intensity, and that it is consistent with any reference data (for example, relative sensitivity factors; see Chapter 5) that may be used in the quantification algorithm. For this to be the case, the spectrometer (comprising the energy analyzer, any associated electron lenses, and the detection system) must be both well-behaved and well-characterised. That is, the signal recorded must be linearly dependent on the true signal intensity, and the relative energy dependence of the overall instrument efficiency must be known. The subject of linearity in electron counting and detection systems is reviewed by Seah and Tossa (1992).

Table 3.1 Reference XPS line energies for copper, silver and gold, traceable to the 1990 S.I. definition of the Volt (Seah, 1989). The figures in brackets represent the one standard deviation uncertainties.

	Binding Energies	
	Al radiation	Mg radiation
Cu 3p	75.14 (0.02)	75.13 (0.02)
Au $4f_{7/2}$	83.98 (0.02)	84.00 (0.01)
Ag $3d_{5/2}$	368.26 (0.02)	368.27 (0.01)
Cu L_3MM	567.96 (0.02)	334.94 (0.01)
Cu $2p_{3/2}$	932.67 (0.02)	932.66 (0.02)
Ag M_4NN	1128.78 (0.02)	895.75 (0.02)

In AES, the linearity of the instrument intensity response at a given energy may be checked relatively easily simply by measuring the intensity either of a peak if the differential mode is used or at a suitable energy on the spectral background in the direct mode, as a function of beam current. The beam current should preferably be monitored using an optimum Faraday cup (Ingram and Seah, 1989) during this exercise. Some count rate will be found beyond which the count rate is no longer linearly dependent on the beam current but falls off to an increasing degree, and this count rate should not be exceeded for quantitative analysis.

It is not quite so straightforward to perform the equivalent experiment in XPS, as there is no readily available convenient way of measuring the photon flux from the X-ray source. Instead, the peak-to-background ratio of, for example, the copper $2p_{3/2}$ peak may be measured as a function of peak count rate while varying the X-ray source power. A count rate will be found beyound which the peak to background ratio falls off. An acceptable loss of signal should be determined, for example 2% of the peak intensity, and data for quantitative analysis should not be accumulated at higher count rates

than that value. An example of such a result is shown in Figure 3.6, where a
5% loss in intensity is observed at 100 kcounts s^{-1}. The loss of intensity
at high count rates in pulse counting systems is usually due to the dead-time
of the amplifier/pulse-shaper cicuit used. In general:

$$n = n_0 \exp (-t/T)$$

[3.10]

where n is the measured couunt rate, n_0 is the true count rate, t is the mean
time between arrival of pulses and T is the electronics' response time, or
dead-time. For the example given in Figure 3.6 the response time is a not
unreasonable value of 500 ns.

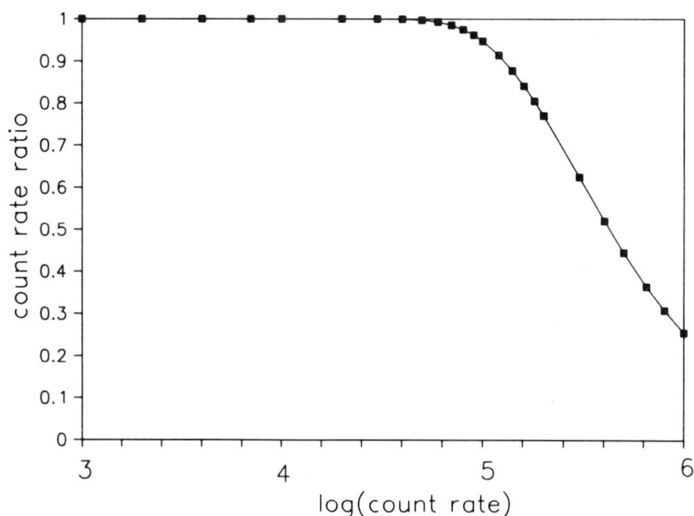

Figure 3.6 A graph of the ratio of count rate on the copper $2p_{3/2}$
photoelectron peak determined with an X-ray source power supply
emission current of 20 mA to that at 5 mA, as a function of the
count rate at 20 mA. The count rate at 20 mA was varied by
adjusting the high voltage applied to the source. The solid line
shows the best fit of the data to equation [3.10].

In general, for any electron spectrometer the measured intensity as a
function of electron kinetic energy, I(E) is related to the true spectrum
n(E) through a spectrometer response function Q(E) (Seah and Smith, 1990):

$$I(E) = Q(E)n(E)$$

[3.11]

where Q(E) is a combination of several terms:

$$Q(E) = H(E)T(E)D(E)F(E)$$

[3.12]

Here, T(E) and D(E) are the energy dependencies of the spectrometer
transmission and electron detection efficiencies; H(E) is a term allowing for
design and manufacturing tolerances, aberrations and stray electric and
magnetic fields; F(E) describes the efficiency of the electronic system
transferring the detector output to its recorded form.

Usually, spectrometers are sufficiently well designed, constructed and set up that H(E) is effectively unity except for low energy electrons, possibly up to around 50 eV kinetic energy. For an instrument in which the multiplier voltage and discriminator threshold are correctly set, the F(E) term makes only a negligible contribution to Q(E). This leaves the transmission and detection terms, which were briefly discussed in the context of analyzer design in Section 3.3 above.

Calculations of electron energy analyzer transmission functions have been published (Seah, 1980) showing, for the constant $\triangle$E mode of hemispherical sector electron energy analyser operation, energy dependencies of $E^{-\frac{1}{2}}$ or E^{-1} may be expected, depending upon the precise operating conditions. Early experimental determinations of the transmission of commercially available analyzers appeared to give results in more or less good agreement

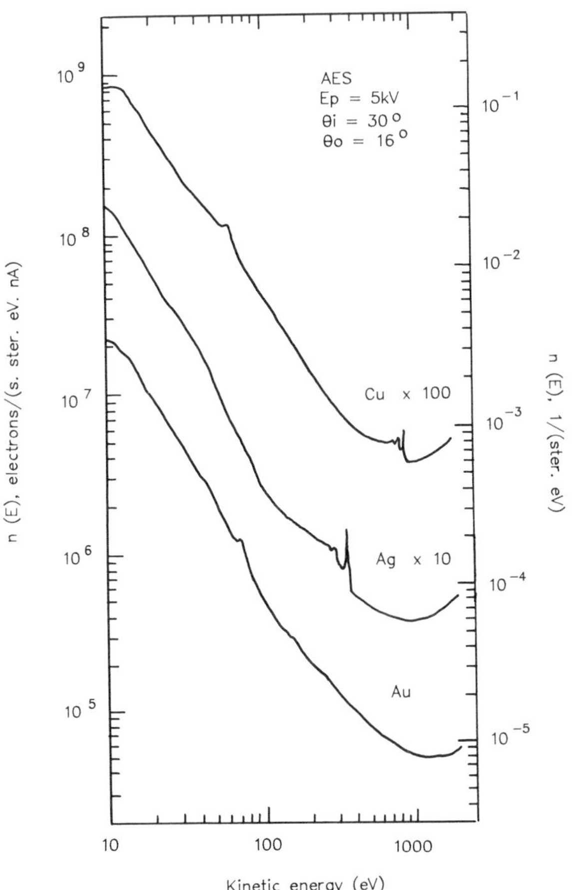

Figure 3.7 Reference spectra for AES as measured at the UK National Physical Laboratory using the Metrology Spectrometer (Smith and Seah, 1990). The spectra are plotted on an absolute intensity axis, and their shapes contain no contributions from the energy dependencies of the transmission or detection efficiencies. The silver and copper spectral intensities are multiplied by factors of 10 and 100 respectively, for clarity.
(reprinted with permission, John Wiley and Sons Ltd.)

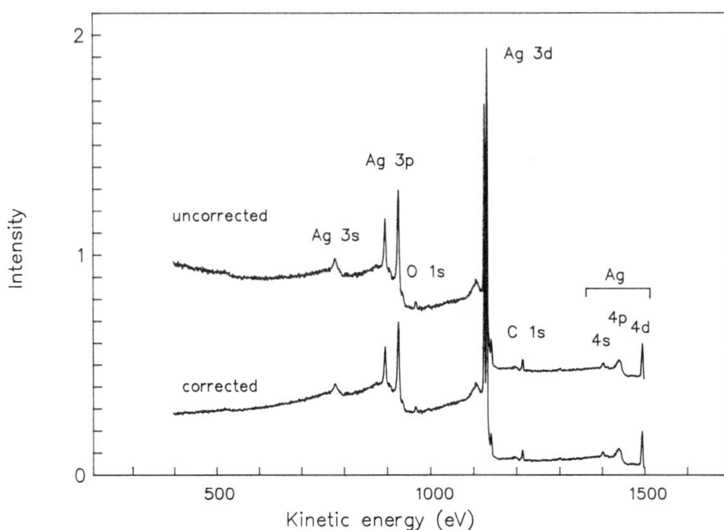

Figure 3.8 XPS spectrum from a contaminated silver surface, shown before and after correction for the energy dependence of the transmission of the electron energy analyzer used.

with theory (Cross and Castle, 1981; Hughes and Phillips, 1982; Scharli and Brunner, 1983; van Eenbergen and Brunix, 1984). However, typical analyzer behaviour is now known to be more complex than was originally thought, particularly where electron transport lenses are used, and complex ray-tracing methods are required for a full understanding of analyzer transmission characteristics (Osterwalder et al, 1989). Consequently, it is better to rely on an experimental determination of T(E) obtained by comparison with reference spectra.

Considerable effort has been expended at the UK NPL on the generation and verification of reference spectra for XPS and AES for the purpose of analyzer response characterisation. This has culminated in the determination of reference spectra for both techniques in which the energy-dependence of the instrument function has been completely removed, leaving the true spectral distribution (Seah and Smith, 1990). In AES it was found necessary to define standard operating conditions of a 5 kV beam energy incident at 30° to the sample normal for electrons detected at emission angles up to 65° from the normal to the sample surface. Figure 3.7 shows the reference spectra for sputter-cleaned copper, silver and gold surfaces, with an absolute intensity scale in terms of electrons emitted per unit of incident flux for a given geometry (Smith and Seah, 1990).

The effect of properly calibrating the intensity scale of an XPS spectrometer is illustrated in Figure 3.8. The figure shows the XPS spectrum of a contaminated silver sample before and after correction for the transmission of the analyzer. The correction was carried out by measuring spectra from clean copper, silver and gold, and deriving the energy dependence of the analyzer transmission from the comparison of these with the standard reference spectra for XPS, using unmonochromatised Al Kα radiation from the literature (Seah and Smith, 1990). The analyzer used for this comparison was of a simple design, with no input lens, whose transmission is expected to vary with a kinetic energy dependence between $1/\sqrt{E}$ and $1/E$. The uncorrected spectrum falls off much more steeply with energy than the corrected spectrum, demonstrating this effect.

For XPS and AES, the NPL reference spectra have been verified by inter-laboratory comparisons, with repeatabilities of the order of a few percentage points for calibrated instruments. This compares favourably with the order of magnitude errors found in the earlier ASTM E-42 sponsored work.

Recently, various workers have attempted to determine transmission functions independently (Kurbatov and Zaporozchenko, 1991; Smith, 1991; Carrazza and Leon, 1991). Where comparisons are possible, the results appear to be in agreement with the NPL measurements.

3.4 Radiation Damage

In surface analysis by electron spectroscopy, it is important to consider the possibility of modification or damage induced in the surface region of the specimen as a consequence of exposure to the interrogating electron or X-ray beam. The susceptibility of a particular surface to radiation damage limits the incident beam flux densities that may be used, and this in turn has consequences for the sensitivities and spatial resolut-ion that may be achieved.

The possibility of relatively high current densities in AES makes the technique particularly prone to radiation damage, and beam induced artefacts are quite common, especially on insulating samples. For example, many oxides readily decompose under electron radiation, giving a peak in the spectrum due to the metal in the metallic state in addition to the expected metal peak at the energy typical of the metal oxide (Holloway, 1976). This electron beam decomposition effect represents an additional hazard for correct spectral interpretation (Burstein, 1980), and has severely restricted the application of AES in the field of catalysis (Barr, 1983). An assessment of the damaging effects of the electron beam in AES is given by Pantano and Madey (1981), who provide a useful overview and a ranking of the differing susceptibilities to electron beam damage of a range of materials.

Problems of beam damage are generally held to be less severe in the case of XPS, although on polymer surfaces prolonged exposure to the X-ray beam has been observed to produce visual evidence of damage in the form of discolour-ation. Chemical changes include depolymersation, emission of light molecules, and cross-linking of the polymer chains (Clark, 1977). Chlorine-or flourine-containing polymers (especially polyvinylchloride - PVC) seem susceptible to this kind of damage. Partial reduction of oxides, similar to but less pronounced than for electron beam irradiation, has also been observed.

The relatively benign nature of XPS compared to AES may, in part, be simply due to the lower flux densities that are used (Cazaux, 1985). Therefore, a true comparison between the techniques should be carried out for the same spatial resolution. If that is done, then it appears that the main advantage in using XPS is in its greater sensitivity. That is, for a given incident flux density, the peak to background ratio of a typical peak may be around an order of magnitude greater than in AES, and so the counting time required to build up statistically signficant data is correspondingly reduced. In addition, when the dominant mechanism causing surface damage is ionisation by the primary beam (as opposed to secondary or backscattered electrons within the sample), the damage cross-section of XPS may be up to two orders of magnitude less than for AES. Thus, for a given element being analyzed at a given spatial resolution the susceptibility to surface damage could typically be around two orders of magnitude less for XPS than for AES.

However, at present, the areas of analysis of the techniques in routine operation are separated by perhaps six orders of magnitude and the dominant effect causing the apparently less aggressive nature of XPS is simply the much lower flux densities in use. Radiation damage will certainly become more of an issue in XPS as the trend towards higher resolution continues, as discussed in section 3.6 below.

3.5 Electrostatic Charging

In Chapter 2, examples were given of X-ray photoemission and Auger electron emission from elements in different local chemical environments. In both cases shifts in peak energies were observed, together with some changes in peak structure. In order to make correct assignments of chemical states, accurate measures of the energies of the peaks are needed. This requires that the energy scale of the spectrometer is properly calibrated using known atomic standards for reference line energies and following recommended procedures as discussed in section 3.3.4 above. A further requirement is that the problem of sample charging be adequately dealt with.

Charging arises when insulating samples accumulate or lose electronic charge as a consequence of the electron spectroscopic measurement. In AES, an insulating sample will acquire electrons from the primary beam which can not be conducted away. If the material of the sample has a secondary emission coefficient of less than unity, then insufficient electrons will be lost to compensate and a negative charge will build up. Similarly, a large value of the secondary emission coefficient will cause positive charging. For a positively charging surface, application of a positive specimen bias voltage will attract back most of the low energy secondary electrons, and with careful adjustment a balance may be obtained. Negative specimen charges are more difficult to deal with, although optimum values of the electron beam energy and angle of incidence can often be found to minimise the problem. In any case, small values of electrostatic charge on the sample result in corresponding shifts of peak energies across the spectral range and need not cause a failure of the experiment provided the effect is understood. Serious charging on a highly insulating sample can result in catastrophic distortion of the spectrum, possibly coupled with deflection of the primary beam, rendering AES useless as a tool for probing such a surface. The background to charging and charge compensation in the analysis of insulators by AES is given by Hofmann (1992).

In XPS, the tendency is for the sample to become positively charged as a result of the loss of electrons in the photoemission process. However, the standard twin-anode X-ray source also generates a significant number of low energy electrons as a result of photoemission from the aluminium window. Some of these may be attracted to the positively charging sample, and may partially neutralise its charge. If a monochromatic X-ray source is used then there are far fewer stray electrons present in the system and a low-energy "flood" beam of electrons may be introduced to help overcome the problem. Despite this, the analyst is often left with a spectrum in which all the peaks appear to be systematically shifted by a few electron volts from their expected energies. In such cases, it is common practice to identify the component of the carbon 1s peak which is due to hydrocarbon contamination (almost always present), note its energy difference from the expected value, and apply the same shift to all other peaks in the spectrum before making any chemical state assignments. This is good practice, however, different authorities differ on the exact value of the carbon 1s hydrocarbon binding energy that should be used. A figure of 284.7 eV is given in the Perkin-Elmer Handbook (Wagner et al, 1979), the NIST data base (National Institute for Standards and Technology, 1988) uses 284.8 eV, and various instrument manufacturer's data systems use values of 285.0 eV. The

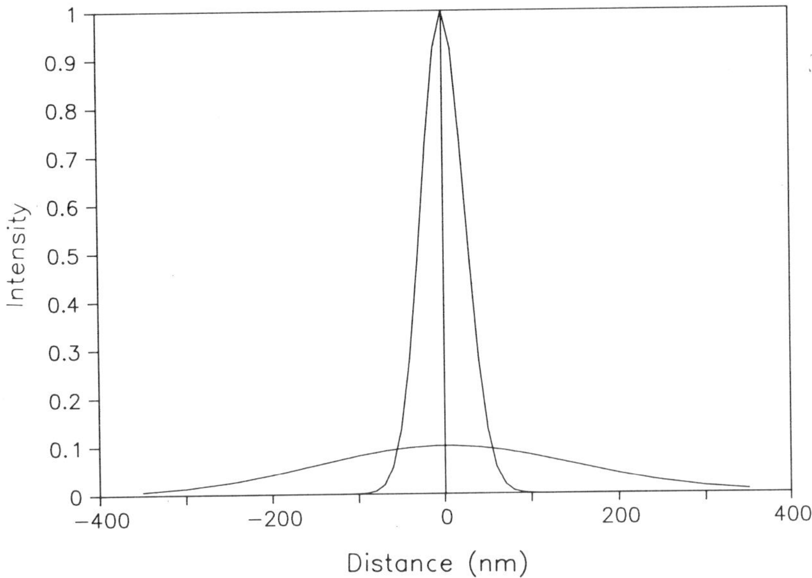

Figure 3.9 A schematic representation of the effect of backscattering of
primary electrons and the subsequent Auger electron generation on
the attainable spatial resolution in AES. The narrow peak of
intensity arises from Auger electrons excited by the primary
beam. This is superimposed on a broader halo of intensity arising
from Auger electrons generated by backscattered primaries.

reliability of this method has been reviewed by Swift (1982). Current best
practice appears to be to use a value of 285.0 eV with an associated
uncertainty of 0.2 eV.

Often, simple charge referencing as described above is inadequate. This
is especially true when an inhomogeneous sample, with surface regions of
varying conductivity, is analysed. In this case, different parts of the
surface may charge to different degrees, resulting in the peak broadening or
splitting that is characteristic of differential charging. Approximate
neutralisation by stray secondary electrons originating from the X-ray
source cannot compensate for this effect and the use of a controlled charge
neutralisation device is necessary. The use of a low energy electron flood
gun has already been mentioned. In some spectrometers this is used in
conjunction with a fine nickel wire grid held at earth potential just above
the sample surface (Barth et al, 1988) and the method is found to be
extremely effective. The various methods available for charge compensation
in XPS are reviewed by Cros (1992).

3.6 Microanalysis and Spatial Resolution in AES and XPS

The essence of microanalysis is the achievement of good spatial
resolution, and, for the surface sensitive electron spectroscopies, this
includes both lateral resolution on the sample surface and resolution in
depth. For AES and XPS, the depth resolution is governed by the escape depth
of the emitted electrons. From Figure 2.1, this is typically in the range of

a few monolayers. The lateral resolution may be governed by the physics of
the process or by the experimental arrangement, itself involving either the
finite probe size, the area selected by the input optics of the analyzer or
the imaging properties of the system. The different factors operating for
AES and XPS are discussed separately below.

3.6.1 Spatial Resolution in AES

For practical purposes in most routine analyses, the lateral resolution
in AES is limited by the diameter of the incoming electron beam. However,
for very small probe sizes, the resolution may be degraded by Auger emission
from atoms that have been excited by backscattered primary electrons. Figure
3.9 shows this effect schematically. The result is a halo of emission of
both backscattered and Auger electrons around the point of incidence,
typically of the order of 100 nm diameter, depending upon the material under
investigation (Tokutaka et al, 1987). This corresponds to the minimum
achievable resolution observed in a large number of experimental studies,
although commercial systems with spot sizes in the region of 50 nm are now
available.

As a consequence of the backscattered electron contribution, the
resolution function for very small beam widths has two components - a narrow
peak of approximately Gaussian shape due to the finely focussed incident
beam, superimposed on the low intensity but much broader distribution of the
halo. It is clear that, when scanning over a very small area, the
backscattered contribution is effectively uniform and the resolution function
is dominated by the incident beam profile. Using a 5 nm beam diameter,
Janssen and Venables (1978) have presented results which imply a resolution
of   20 nm if the usual criterion of a 16% to 84% change in signal over an
edge is adopted, despite the presence of a halo due to Auger generation by
backscattered electrons extending over 100 nm from the edge.

In many cases, the specimen itself can contribute to the apparent
resolution degradation, or to the appearance of artefacts in the spectra.
Analysis of microelectronic devices is particularly susceptible to the kind
of effects shown in Figure 3.10 (El-Gomati et al, 1987). Here, the intensity
of the silicon Auger peak is monitored as the incident beam is scanned over
the aluminium layer shown in the figure. On the left hand side of the
aluminium layer the silicon signal is prematurely attenuated by shadowing
whereas in the right hand side an enhancement is observed, caused by
generation of high energy scattered electrons in the aluminium layer which
then emerge and impinge upon the silicon, causing Auger emission. In Auger
imaging this would give the misleading result of an unexpected dark line in
the silicon map on one side of the aluminium island, and a corresponding
unexpected bright line on the other side of the island, from which incorrect
conclusions could easily be drawn were the effect not understood.

In order to produce sub-micron diameter probe beams, high accelerating
voltages are necessary. However, the amount of backscattering also tends to
increase with the kinetic energy of the incident beam, therefore some design
compromises must be made. The best spot sizes are produced with beam
energies of up to 100 keV such as may be encountered in a UHV scanning
electron microscope. However, it is generally accepted that as low a beam
energy as possible should be used, consistent with obtaining the desired spot
size. In addition, for optimal detection, it is desirable to maximize the
(energy dependent) peak to background ratio of the signal of interest. Such
considerations may lead to choices of beam energies in the 10 - 30 keV range
(Venables et al, 1986), although there is more information on quantification
generally available for 5 keV beams. Under certain circumstances, there may
be advantages in using very high (100 keV) beam energies, particularly for

the analysis of small precipitates, as a higher degree of ionization may be
achieved (Cazaux, 1987).

A danger that is always present when very small probe beams are used is
the exposure of regions of the sample to high current densities, and the
consequent risk of beam-induced specimen damage, as discussed in section 3.4
above.

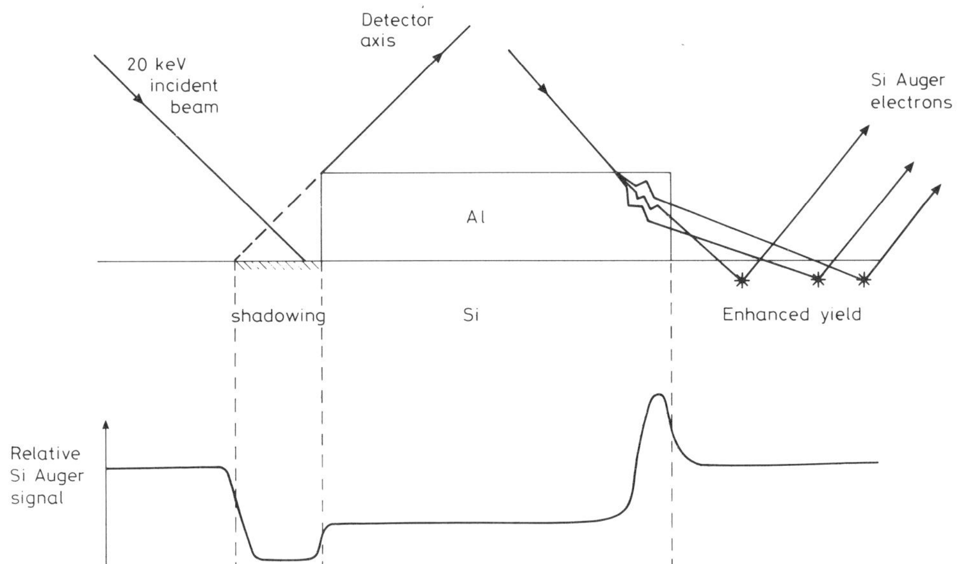

Figure 3.10 Illustrating the artefacts that can arise during AES analysis of
samples with a high degree of surface topography. The figure
shows schematically the apparent reduction of intensity at one
side of an island of aluminium on a silicon substrate and the
apparent enhancement of signal on the other side as the incident
electron beam is scanned from left to right across the structure
(after El-Gomati et al, 1987).

3.6.2 Spatial Resolution in XPS

For obvious practical reasons, beams of X-rays cannot be simply focussed
and scanned in the same way as for electrons. This imposes restrictions on
the image-forming capabilities of XPS instruments, and has lead to the
development of novel experimental designs. Several of these are rather
exotic and not suitable for incorporation into conventional multi-technique
surface analysis apparatus. Examples include the photoelectron spectromicro-
scope (Beamson et al, 1981), the photoemission microscope (Griffith, 1986)
and the use of X-ray optical methods such as zone plates (Kirz and Rarback,
1985). Practical methods of small-spot XPS and XPS imaging usually rely on
restriction of the source size or the analysed area, often in conjunction
with the use of a position sensitive detector (Seah and Smith, 1988).

Restriction of the source size on the sample can be achieved simply by
placing a suitable aperture or collimator between the X-ray source and the
sample, however, this would result in several orders of magnitude loss in
signal. A better method is to use the focussing properties of a crystal

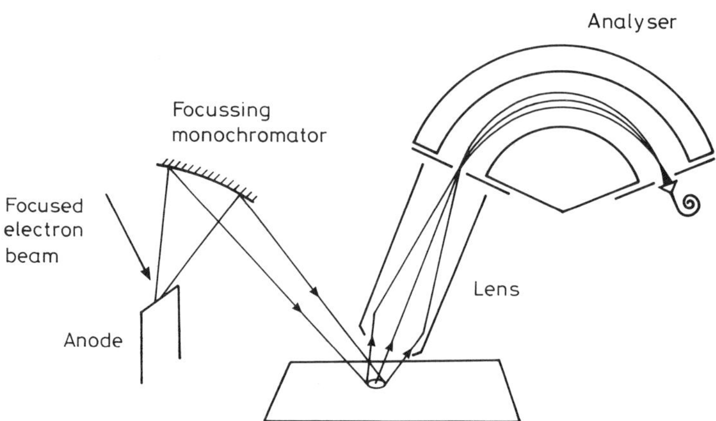

Figure 3.11 Achievement of spatial resolution in XPS using a focussed
electron beam to excite X-rays which are subsequently collected
and focussed onto the sample using a curved crystal
monochromator.

monochromator, which also gives the advantages of better energy resolution
and higher signal to background ratios in the spectra. The normal design of
monochromator, optimised for high signal and good energy resolution, uses a
curved crystal to give a line focus of X-rays on the sample surface from an
electron-irradiated anode. An advance on this is to use a crystal curved in
two planes to give a point focus on the specimen. Control of the X-ray spot
size may be obtained through a variable focus of the electron source
impinging upon the anode. With good design, X-ray beam sizes contollable
over a range of around 100 μm to 1 mm may be achieved (Chaney, 1987). The
useful throughput of such a system may be optimised by the use of an electron
spectrometer whose acceptance area on the sample is equal to, and aligned
with, the small illuminated spot. This involves the use of an electron lens
of low aberration between the sample and the analyzer, with magnification
such that the image of the illuminated spot at the input slit is matched to
the size of that slit. Ideally, this lens would be of high acceptance angle
for maximum signal collection, although optional restrictions on the solid
angle of collection may be necessary if angular-dependent work is to be
undertaken. Use of a parallel data collection system involving a position
sensitive detector further increases the data collection efficiency. Such
small-spot XPS systems, as shown schematically in Figure 3.11, are
commercially available in dedicated XPS instrumentation and have been
successfully used in many investigations.

A technique for producing a small source of X-rays has been developed
which involves irradiating the aluminium-coated rear of a thin sample with an
electron beam (Cazaux, 1984). The electron beam generates aluminium Kα X-
rays which propogate through the sample towards its front surface, generating
photoelectrons as they travel. Photoelectrons produced within the usual
escape depth of the front surface are then analysed and detected by a
conventional electron spectrometer. The technique is occasionally referred
to as transmission-XPS and is outlined in Figure 3.12. This method involves
special sample preparation techniques but has the advantage that XPS images
may be produced. By tuning the analyser to a peak in the XPS spectrum of the
element of interest, and using this peak intensity to modify the intensity of

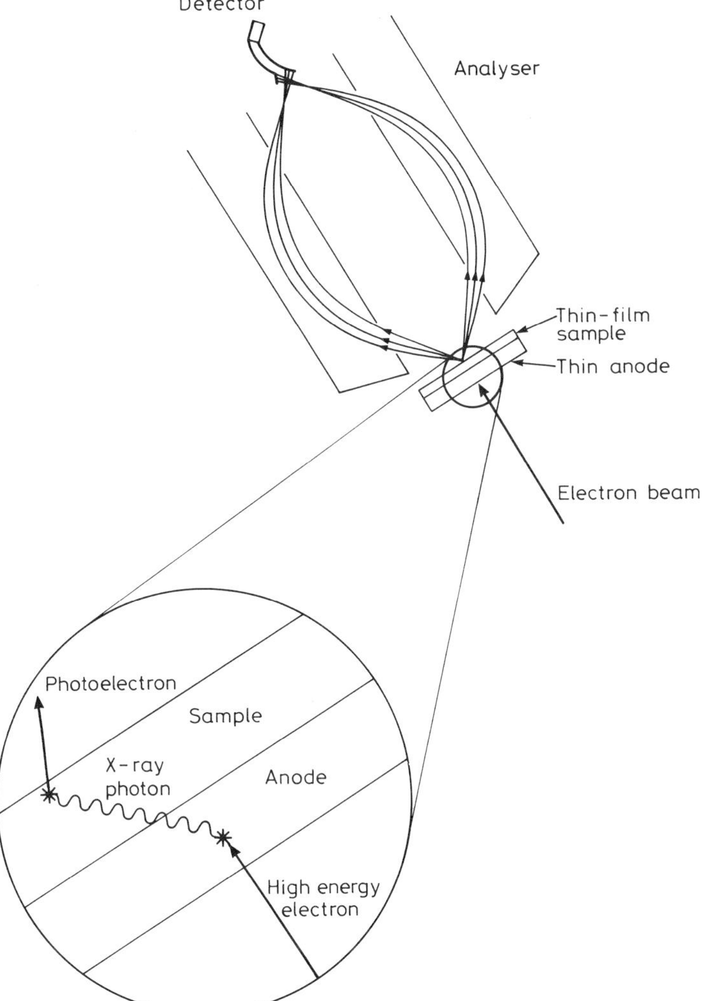

Figure 3.12 A method of achieving spatial resolution in XPS due to Cazaux (1984) in which a thin X-ray anode is deposited on the rear of the sample. The anode is addressed by a focussed electron beam which gives rise to a small X-ray generation volume immediately below the sample.

a display scanned synchronously with the electron beam, images may be built up whose brightness depends upon concentration variations across the specimen surface. This is thus the XPS analogue of the scanning Auger microprobe.

A method of generating XPS images in conventional apparatus without using the transmission geometry has been developed by Gurker et al (1983). Here, the sample is flooded with X-rays and a line source is selected using a narrow input slit to the analyser. Electrons of the chosen peak energy are focussed by the spherical sector analyser to the output slit, where they are detected using a position sensitive detector. The two-dimensional position sensitive detector records spectral information along the dispersive direction normal to the major axis of the input slit, and spatial information along the line of the input slit. The image is formed by recording the

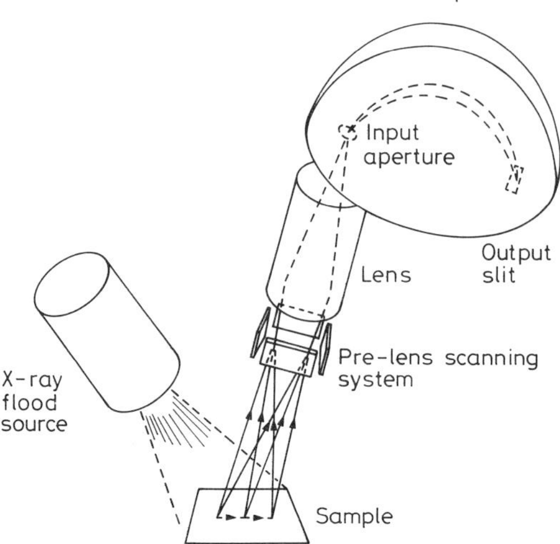

Figure 3.13 A simple method of XPS imaging using a conventional HSA
instrument fitted with deflection plates between the sample and
the input lens which are used to scan the image of the input slit
across the sample surface (Smith and Seah, 1987).

changes in intensity at the output slit as the sample is mechanically scanned
along the dispersion direction of the analyser. A much more sophisticated
instrument, but essentially employing the same principle, has been described
by Gelius et al (1990). Again, a position sensitive detector is used,
however, a focussed monochromator of high flux density defines the line of
interrogation on the sample surface. The images produced by the system
consist of plots of spectral information in a chosen energy window against
lateral position on the specimen surface. Resolution of around 10 μm is
possible, although the instrument is designed for very high sensitivity XPS
measurements rather than specifically for imaging. The practical use of this
instrument is discussed by Beamson et al (1990).

In principle, any CHA-based system with an input lens can be used in the
small spot mode by selecting the highest lens magnification and smallest
input aperture available. Such a facility is extremely useful for routine
analytical applications. Using this kind of selected area system, XPS images
could, in principle, be obtained by performing a mechanical x-y raster scan
using the specimen manipulator and noting the intensity changes. However,
the method is clumsy and a much more attractive variant is to incorporate
electrostatic deflection plates between the sample and the lens (Smith and
Seah,1987). With the sample flooded with X-rays, the analyzer and lens are
used in the selected area mode and the deflection plates enable the image of
the input aperture (the selected area) to be raster scanned across the sample
surface. Synchronous modulation of the intensity of a display screen allows
an XPS image to be built up. This system gives spatial resolution in the 100
μm to 200 μm range for the price of a minor modification to an existing
spectrometer. An outline of the arrangement is shown in figure 3.13 and, as
an example of its application an XPS image of the solder pads of a surface
mounting electronic package is given in figure 3.14. To give an indication
of scale, the gold solder pads, imaged with the Au $4f_{7/2}$ line, are 0.635 mm
along their minor axis.

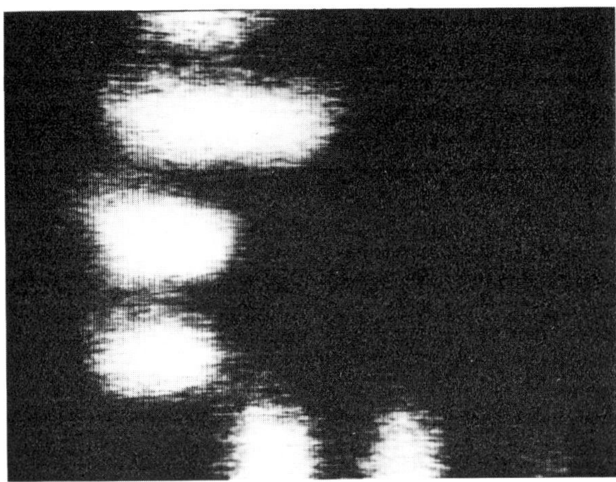

Figure 3.14 An XPS image of the solder pads on a surface mounting electronic device package, acquired using the scanning method shown in Figure 3.13.

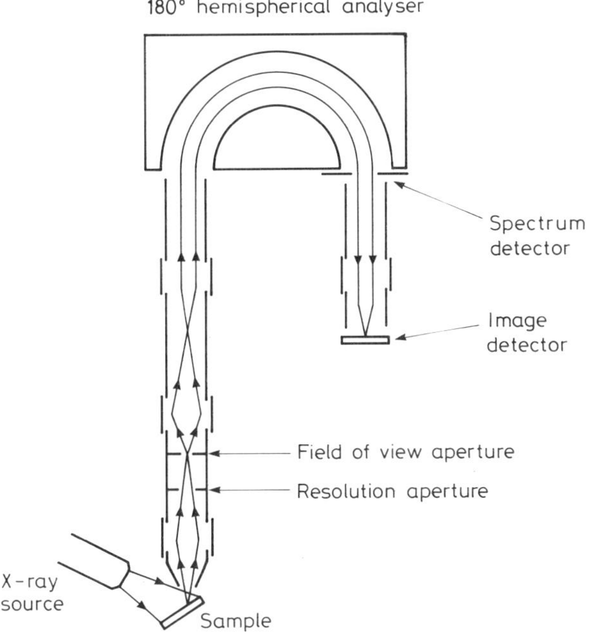

Figure 3.15 Schematic diagram of the electron optical arrangement of a parallel imaging XPS spectrometer (Coxon et al, 1990) (reprinted with permission, Fisons Instruments, Surface Science Division)

Both mechanical scanning of the stage and electrostatic scanning of the selected area can now be found on commercially available instruments. An interesting variant of the electrostatic scanning method, also found on a commercial instrument, is the inclusion of a magnetic immersion lens below the sample position (Drummond et al, 1991). This appears to fulfil the dual purpose of improving the spatial resolution available, and increasing the signal intensity for a given spatial resolution. It should be noted that these methods may not be suitable for radiation sensitive materials, as the whole of the sample is continuously irradiated during image acquisition, whereas with systems using a focussed monochromator only that part of the sample currently contributing to the image is subject to the X-radiation.

In optical microscopy, the entire image is collected simultaneously, in parallel. Recent progress has resulted in the development of a novel instrument which performs the electrostatic analogue of optical microcopy, whilst retaining the energy selecting properties of a hemispherical analyzer, to give spatial resolution of the order of 10 μm (Coxon et al, 1990). An outline of the electron-optical scheme is shown in figure 3.15. The input lens acts in such a way that the position of origin of electrons on the sample surface is converted to angular information which is retained during energy analysis by the CHA. Angular information is converted back to positional information by the output lens, where an image is formed on the surface of a position-sensitive detector. Electrons in the image plane may also be re-focussed onto a normal electron multiplier, therefore enabling spectra to be recorded from very small (<10 μm) areas on the sample. This represents at least an order of magnitude improvement over existing capabilities, and the XPS technique may now be applied to a range of surface microanalysis problems for which it was previously thought unsuitable. The application of such an instrument to the study of tribological interactions between polymer and metal surfaces has been reported by Fletcher et al (1992).

DATA PROCESSING FOR AES AND XPS

4.1 Introduction

In common with the vast majority of current analytical instrumentation, almost all surface analysis instruments in use in industrial or analytical laboratories are under computer control and data are acquired and stored in digital form. The result is an increased output and a saving of skilled operator time. However, there is a risk, particularly in fully automated systems, that the operator loses touch with what is actually happening in the machine and may fail to recognise erroneous or meaningless results.

In analytical electron spectroscopy the purpose of computer control of the spectrometer is to allow series of experiments to be set up and run automatically, with post-acquisition processing of the resulting data. A typical acquisition strategy may consist of preprogrammed multiple scans over selected regions of the spectrum. More sophisticated instruments allow automatic sample change-over, multi-point analyses and variable angle measurements under computer control. This enables a series of experiments to run overnight with the results of the analysis, possibly including a set of predefined data processing operations, presented to the operator on returning the following morning. A problem with this kind of equipment is that it can become virtually impossible to keep up with the volume of data generated, and the throughput of samples to be analyzed is limited by the data interpretation time rather than the capacity of the instrument.

A schematic diagram of a possible configuration of a simple electron spectroscopy data acquisition and processing system is shown in Figure 4.1. The computer has direct control, through an appropriate interface, of the primary excitation source and of the analysis and detection of electrons, and the means are provided for data processing operations and output either as hard copy or in digital form possibly for transfer to a central processing and archiving facility or to a laboratory information management system. The types of computer used typically range from a basic personal computer through various degrees of sophistication up to fully integrated work-station based systems. The simpler systems allow single task operation; either control of the spectrometer and data acquisition, or post acquisition processing. More complex and powerful machines will support multi-task and multi-user operations, with foreground/background facilities so that previously-acquired data may be processed while further data acquisition is in progress. The features of a number of electron spectroscopy data systems are described

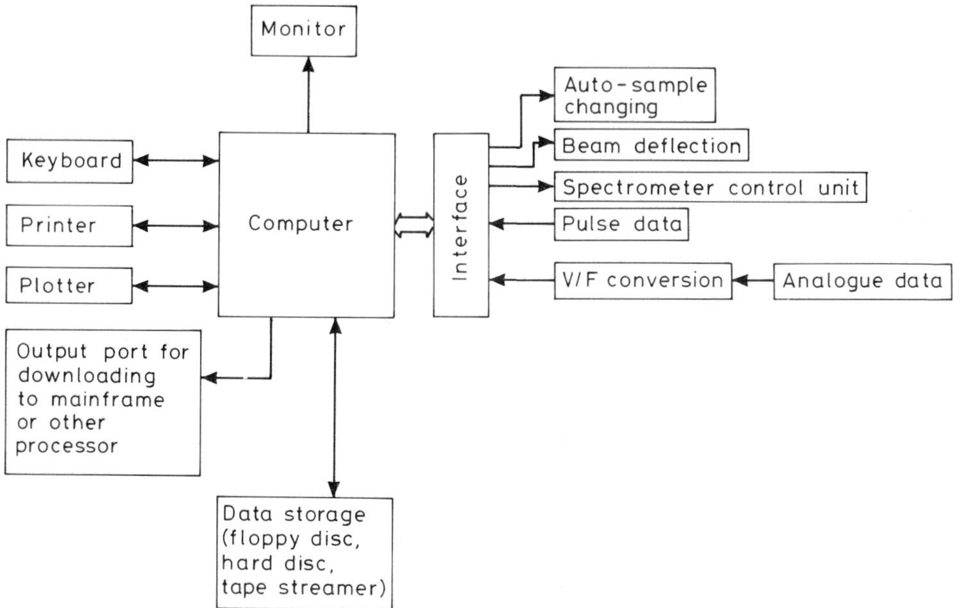

Figure 4.1 Simplified schematic arrangement of a spectrometer control system
with data acquisition and processing.

in the literature (e.g. Fitzgerald et al, 1992; Desimoni and Biader Ceipidor,
1991).

4.2 Data Acquisition and Display

All electron spectrometers require a linear ramp voltage, with an
appropriate amount of amplification, to control the programmable high voltage
power supplies that drive the electron optical components providing energy
dispersion. This is true even where parallel collection is used, as the
position-sensitive devices that are usually used can only cover a part of the
spectral region of interest. Use of a 16-bit digital to analogue converter
to generate this ramp voltage will, in principle, allow a spectrum to be
scanned in steps of 0.05 eV over a 3 keV range, covering most of the Auger
peaks of interest in AES. For conventional XPS with either Mg or Al
radiation the scan width may be reduced to a maximum of, for example, 1.5
keV, allowing an improved resolution or a reduction in the accuracy of the
DAC. As the spectrum is scanned, pulses from the electron detector are
shaped and amplified before being accumulated in the computer memory to build
up the spectrum sequentially channel by channel.

In the data acquisition mode, the computer control system will allow the
start and end points of the spectrum to be set up, together with scan
parameters such as the width of each channel in electron volts, the dwell
time per channel and the number of scans to be made. It can be advantageous
to select a large number of scans with a short dwell time rather than a
single scan with a large dwell time per channel but the same total data
acquisition time, since this will help to average out any drifts in the
primary beam current or incident flux that may occur. However, the maximum
slew rate of the high voltage supplies driving the electron energy analyser
is likely to be a limiting factor in the scan rates that can be achieved.
As an alternative to fixing the scan time, some systems allow continuous

scanning until a pre-determined value of signal to noise ratio is obtained, enabling a uniform data quality to be achieved. However, this has the disadvantage that the instrument can spend a great deal of time on regions of the spectrum where there are no peaks and hence low count rate, with relatively little time spent on the peaks of interest. Most data systems will allow regions to be set up enabling several spectra to be acquired from one sample during one experimental run. This may involve, for example, a wide-scan XPS spectrum followed by detailed scans over individual peaks at higher energy resolution. If depth profiling using an ion beam is carried out (see Chapter 6), the system should be able to allow the ion beam to erode the sample for a fixed time before repeating the spectra and storing the results in memory. Automatic peak area measurements and presentation of the data as intensity against time of exposure to the beam should also be possible, as the depth profile develops in real time. For imaging AES under computer control, the system will direct the electron beam to the next pixel to be interrogated. Alternatively, an automatic sample exchange or a change in sample angle may take place, followed by further data acqusition.

If the incoming data is in the form of an image, typically from a scanning AES experiment, it will be stored pixel by pixel probably in a block of specially configured computer memory known as a frame-store. Typical image sizes may vary from 64 x 64 8-bit pixels up to 512 x 512 or greater. The frame-store circumvents the difficulties associated with reading such large quantities of data from conventional memory. Furthermore, each image frame may be supplemented by one or more overlay planes into which graphical and alphanumeric data associated with the image may be written. This information can then be superimposed on the image during display without affecting the quality of the stored data.

Once the data have been acquired and stored, a variety of processing operations may be carried out. It will at some point be necessary to display the spectra in the form of signal intensity in counts per channel against energy in electron volts. In AES, the energy axis should be in electron kinetic energy. For XPS, kinetic energy was originally used during the initial developmental stages of the technique (as the analyzers work in kinetic energy) but it is much more convenient from the point of view of spectral interpretation to arrange the display so that the horizontal axis appears in units of electron binding energy. This requires that the computer system "knows" the spectrometer work function and the X-ray photon energy, either on the operator's instruction or through direct switching from the X-ray source power supply in the case of a twin or multiple anode system.

Multiple display of spectra in a montage form can be helpful in bringing out trends in a sequence of data, for example in an angle-resolved XPS experiment or in depth profiling by argon-ion erosion. For the display of concentration maps in AES, there is the problem that surface topography can influence the intensities recorded and a simple display of Auger peak heights across the specimen can give extremely misleading results. For this reason, it is customary to include some kind of topographic correction in the display software. The use of $(P-B)/(P+B)$, where P is the intensity of the peak of interest and B is the background intensity at some suitable point on the high kinetic energy scale of the peak, has been found to be successful in removing most of the topographical contribution to the measured change in peak height (Browning et al, 1985). For any of these display options to be meaningful, the measured peak heights or, more reliably, areas, must be quantified to give concentrations in the surface region using a reliable and verified routine. This is not necessarily straightforward and, although one of many possible data processing options, quantification can be a complex topic. It is covered separately in Chapter 5.

As well as spectral display, the operator may require further processing
of the data before a quantification is attempted, or some other final result
is produced. In analytical electron spectroscopy, the peak shapes may
contain a great deal of information, for example when looking for subtle
chemical state effects in XPS, and the operator may be trying to extract that
information from a small and correspondingly noisy peak superimposed on a
large energy-dependent background. This can be a taxing problem requiring
the application of a range of data processing operations, including
smoothing, spectral comparison such as addition, subtraction or division,
curve fitting using mixed Guassian and Lorentzian line-shapes, background
subtraction, differentiation and integration, satellite subtraction (in the
case of XPS), and deconvolution. The principles of these processes are
discussed below.

4.3 Smoothing

As a first step towards extraction of meaning from data containing an
element of random noise, it is normal to carry out some kind of smoothing
operation. The objective of smoothing is to remove the high frequency noise
components while retaining, undistorted, all the information concerning peak
heights and shapes contained within the original data. There are two basic
approaches to smoothing, namely use of a convolutional smoothing function or
application of frequency filtering by Fourier analysis. Many commercially
available data systems implement the first method as it is intuitively easier
to visualise the way the algorithms work, especially if a simple multi-point
running smooth (i.e. convolution with a "top hat" function) is used.

After the multi-point running smooth, the most popular convolutional
routine is the cubic/quadratic function of Savitzky and Golay (Savitzky and
Golay, 1964; Steiner et al, 1972), although a simple Gaussian smooth may be
used as an alternative. The Gaussian smooth has the advantage that it
simulates directly the resolution degradation experienced when opening up the
slits on the spectrometer (which may typically have an approximately Gaussian
instrumental lineshape), and therefore its effects are readily appreciated by
the operator. Multiple passes of smoothing functions are sometimes used in
attempts to further improve the quality of noisy data, however, work on
optimal smoothing algorithms has shown that multiple passes of simple
functions give results which tend towards a Gaussian function and that, in
terms of computing time, a single pass of the equivalent Gaussian is as
effective but much quicker. As a general rule the single pass Savitzky-Golay
function gives a low distortion and may be usefully applied when the number
of points in the smooth is as high as the number of points in the full width
at half maximum of the narrowest peak to be resolved in the spectrum (Seah et
al, 1988).

The alternative to convolutional smoothing is frequency filtering in
Fourier space. This is best carried out using an optimal Wiener filter,
defined by reference to the Fourier transform of the data (Kosarev and
Pantos, 1983), and generally gives results very similar to the Savitzky-Golay
convolutional function.

4.4 Deconvolution

The purpose of deconvolution is to remove broadening due to instrumental
factors, and the methods are usually only applied to XPS when the line width
of the X-ray source and the spectrometer resolution contribute to the
observed width of the spectral features. To be succesful, the method
requires data with a very good initial signal to noise ratio.

The usual approach to spectral deconvolution in XPS is based upon the
van Cittert method (van Cittert, 1931; Jansson et al, 1968) of iteratively

solving the matrix equation representing the convolution of the true spectrum and the instrumental broadening function (Carley and Joyner, 1979). The instrumental function is itself a convolution of the spectrometer energy resolution function and the X-ray line shape, with generally a Gaussian distribution being used for the resolution term together with a combination of two Lorentzian functions to simulate the line shape contribution (Beatham and Orchard, 1976). Although it is usual for the iteration to proceed until the changes in the deconvoluted data following a further iteration cycle are below a given root mean square deviation, care must be taken as, beyond a certain point, noise can begin to build up, particularly in the low intensity parts of the spectrum. To a certain extent this can be suppressed by the use of artificial damping parameters, however, in practice a degree of judgement on the part of the operator is often required in determining the optimum point at which to terminate the procedure. In cases like this, there is no substitute for experience gained while working on the data system with real results to analyse.

4.5 Background Subtraction —

In the usual case of a small peak on a large background, in either XPS or direct-spectra AES, it will at some stage be necessary to perform a background subtraction if the detail in the peak is to be examined. This will usually be done after smoothing, if applicable, as the improved signal to noise ratio in the data helps reduce the uncertainty in the exact position of the background. The simplest background to subtract is a straight line between two user-defined points on either side of the peak of interest, as illustrated in Figure 4.2. This is prone to errors involved in the operator's subjective judgement of where the two points should be placed. It is also non-physical and can lead to a significant loss in intensity in many cases (Proctor and Sherwood, 1982). However, it is an option on most data systems and, for simple peak shapes on well defined backgrounds, can give acceptable results if used consistently.

An improvement over the straight line background, involving a closer approach to physical reality is given by the Shirly background (D A Shirley, 1972) also shown in Figure 4.2. This background is generated by assuming that each electron acts as a source for scattered electrons at lower energies, and therefore the background at any energy is proportional to the total number of electrons in the photoelectron spectrum with higher energy. That is, after subtraction of a constant background equal to the intensity on the low binding energy side of the peak, a background correction Δn_i is subtracted from each successive channel, going from high to low binding energy, such that

$$\Delta n_i = b \sum_{j=i+1}^{N} n_j \qquad [4.1]$$

Thus, a smooth curve is fitted to the step often seen on the low kinetic energy (high binding energy) side of peaks. The theoretical basis of using such a background is discussed by Jouaiti et al (1992). The method, with the appropriate scaling and normalising routines, is normally available as an option in commercial data systems for electron spectroscopy. It does suffer from the disadvantage that it is not satisfactory where the background has a slope after the peaks, for example in the 2p photoelectron peaks from many transition metals. Bishop (1981) has proposed a simple modification to the method to allow for this, however, empirically determined fitting parameters are required, reducing the flexibility of the method. Modifications involving an offset to account for the effect of the bandgap seen in the displaced step background from insulating samples have also been suggested.

The Shirley method, with its various modifications, is known to have
significant error (Tougaard and Jorgensen, 1985) but it is a commonly applied
technique which can give good results in practice when used with care.

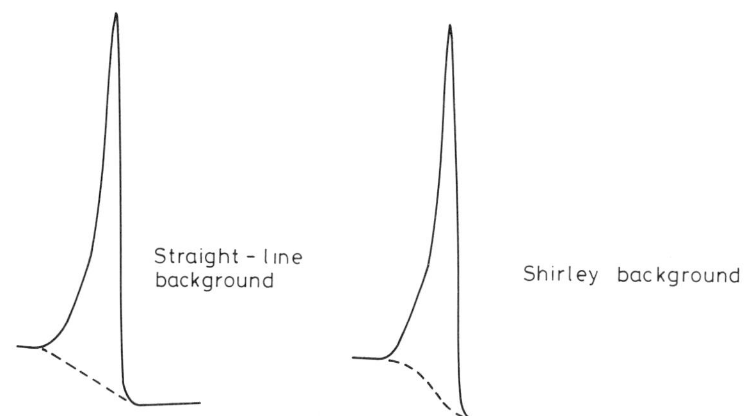

Figure 4.2 An idealized photoelectron peak showing, om the left, the
application of a simple straight line background, and, on the
right, the Shirley background.

An alternative approach to the problem of the background in electron
spectroscopy, based on the physics of the electron inelastic and elastic
scattering processes taking place during photoemission, is the method of
Tougaard (1988). This is particularly attractive as it is widely applicable,
has a sound physical basis, and does not involve arbitrary scaling factors.

Tougaard and Jorgensen (1985) write the primary excitation spectrum $F(E)$
(i.e. the "true" electron distribution which it is desired to obtain by
background subtraction) as

$$F(E) = j(E) - \lambda \int_{E}^{\infty} K(E' - E) \, j(E') \, dE' \qquad [4.2]$$

where $j(E)$ is the measured XPS spectrum (after correction for instrumental
factors), is the appropriate electron attenuation length and $K(E,T)$ is the
probability that an electron of energy E shall loose energy T per unit path
length and energy. This equation, which applies only to homogeneous depth
distributions, ignores elastic scattering of electrons, and assumes that the
cross section for inelastic scattering is the same for all electrons in the
spectrum. Tougaard and Jorgensen (1985) determined $K(E,T)$ from electron
transmission experiments and were then able to deconvolute the background
from a silver specimen excited by Mg Kα X-rays to give a remarkable
agreement between the calculated and experimental backgrounds away from the
spectral peaks. An important result from this work was the identification of
a region of the spectrum extending approximately 50 eV to higher binding
energy above each peak where a significant contribution from primary
electrons was observed. This implies that a simple straight line or Shirley
type fit to the immediate high binding energy side of the peak excludes an
important contribution to the true peak intensity, and also plays a large
part in accounting for the differences between measured and calculated
photoelectron peak intensities discussed in Chapter 5.

Tougaard has pointed out (Tougaard, 1988) that for noble and transition metals the differential inelastic scattering cross sections K(E,T) are sufficiently similar that a so-called "universal" cross section can be used. Furthermore, for other sample types this cross section information can be obtained from an analysis of reflection electron energy loss spectra (REELS) (Chorkendorff and Tougaard, 1987). Such REELS can often be obtained in the XPS spectrometer if it is equipped with a suitable electron gun.

The universal function has the form

$$\lambda(E) \; K(E,T) \simeq BT/(C + T^2)^2 \tag{4.3}$$

where B and C are determined by a least squares fit to experimental data for copper, silver and gold in the energy range 500 - 1500 eV to be B = 2866 eV^2 and C = 1643 eV^2. In later work (Tougaard, 1989) the approach was modified slightly so that equation [4.3] reads:

$$F(E) = j(E) - B_1 \int_E^\infty \frac{E' - E}{[C + (E' - E)^2]^2} \; j(E') \; dE' \tag{4.4}$$

where B_1 is now a fitting parameter. For homogeneous solids, equation [4.4] is valid for the analysis of entire spectra over an energy range of greater than 1000 eV, provided B_1 is fitted such that F(E) is approximately zero on the background away from the peaks. Although initially designed for metals, the universal cross-section appears to be more widely applicable than originally thought and has been successfully applied to semiconductor materials by Bender (1990) and to organic Langmuir-Blodgett films by Sastry et al (1992).

Repoux (1992) has compared background subtraction methods for XPS spectra. For the five elements and one oxide considered, the best agreement with calculated intensities was found when the Tougaard background was used (Repoux, 1992). Overall, it seems that the use of the Tougaard method gives a more physically meaningful background subtraction and is therefore likely to give a higher accuracy than the alternatives when quantification of an unknown sample is required. However, because it requires a wide energy range over which the contribution to the peak intensity is zero it may be difficult to apply to multi-component samples, and the precision can be poor compared to a straight line or Shirley background approach. The method is only now becoming available on commercial electron spectroscopy data systems.

The approach to background subtraction outlined above can also allow information on the relative depth distributions of the elemenets in the sample to be extracted from the data. Because of this, further discussion is left to Chapter 6, where non-destructive methods of obtaining compositional depth profiles are covered.

After background subtraction, most data systems will give the user the option of measuring the peak area. As a measure of intensity for subsequent use in a quantification routine, this is much more reliable than a simple mesaure of peak height and is to be preferred wherever possible.

4.6 Curve Fitting

Often, peaks in XPS or direct mode AES consist of a broad envelope with several major and minor components contributing to its overall shape. These arise as a result of the chemical shifts discussed in Chapter 2, and are an important source of spectral information, particularly in XPS. Correct interpretation of these features requires a model, hopefully giving some idea

of the number of constituent peaks that may be present, together with some
knowledge of the line-shapes and energy widths of the components involved.
The NIST XPS data base is a useful source of such reference data, containing
over 14,000 separate entries on line positions and shifts observed in XPS
(National Institute for Standards and Technology, 1988). There is as yet no
equivalent data base for AES, and the best way to proceed in this case is by
comparison with spectra from known materials ideally measured in the same
instrument as the spectrum for which peak fitting is to be attempted.

Curve synthesis using mixed Gaussian/Lorentzian functions followed by an
optimized procedure for fitting to the experimental data is a good tool for
evaluating the mix of components in a complex line containing contributions
from many chemically shifted states. Its success depends upon the quality of
the experimental data, correct subtraction of the spectral background, and
the accuracy of the operator's first guess at the structure. Usually the
data system will ask the operator to input trial values for peak positions,
heights, widths, and shape (i.e. Gaussian/Lorentzian mixing ratio).
Typically, in XPS the Lorentzian component may be in the region of 10 - 30 %
of the total peak area. Some schemes also allow the use of asymmetric peaks,
although there is a danger of increasing the degrees of freedom too much so
that the physical and chemical meaning of the fit is lost. A good system will
then vary one or more of these parameters until a minimum is obtained in the
root mean square deviation of the synthesised peak shape about the real data.
Some data systems may give a numerical indication of the goodness of fit for
a particular curve synthesis. The reliability of such indicators is
discussed by Cumpson and Seah (1992). It is worth pointing out that the
curve fitting should be carried out on unsmoothed data, and that, if
possible, the level of the background should also be allowed to vary during
the iterations.

Constraints on the fit are often incorporated. For example, it may be
desired to fix certain components of the peak to a given ratio, or to allow
only components within a certain range of peak widths. Typically, for a
well resolved carbon 1s photoelectron peak acquired on a modern instrument,
widths of less than 1 eV may be physically unrealistic, and features greater
than perhaps 1.8 - 2.0 eV may indicate the presence of a further component.
With the best of very high resolution XPS equipment, widths as low as 0.3 eV
may be necessary for separating, for example, the Si 2p doublet. However,
other photoelectron peaks such as the O 1s obtained from organic samples
frequently seem broad even when acquired at the highest resolution, and
features narrower than, say, 1.5 eV are seldom justified by spectroscopic
evidence alone.

Following the initial estimate of the structure, one of a variety of
least-squares fitting routines may be used to refine this estimate until a
close match with the experimental result is obtained. Possibly the most
appropriate method for XPS data is the damped non-linear least squares
technique (Hughes and Sexton, 1988), although it would be unusual to find
this level of sophistication on a commercial data-system. Advanced methods
such as maximum entropy data processing have also been used although, again,
this is uncommon. Instead, the user may be able to proceed manually,
altering the peak parameters in a systematic manner and either making a
subjective judgement or using the computer display of the residuals to decide
on the goodness of fit.

If a collection of spectra all with a different mix of similar chemical
states is available then it may be possible to make use of factor analysis to
determine the number and character of the principle peaks present, reducing
the reliance on the operator's initial guess. Reference data is, of course,
still required to assign the components accurately to the chemical structures

present in the sample. This is also true for the Kalman filter approach to curve resolving complex spectral features (Malitesta and Rotunno, 1991).

Accuracy in peak fitting or curve resolving using the methods described above relies upon the accuracy of the background subtraction used. As discussed in the previous section, this can sometimes be difficult to acheive. A method of avoiding this has been described by Mattogno and Righini (1991) which operates on the first derivative of the XPS spectrum. The approach appears promising but is unlikely to come into common use for some time.

4.7 Difference Spectra

Once the operator has achieved good quality data and has fully identified and analyzed the various peaks contributing to the spectrum, then it may be necessary to compare it with data from similar samples in order to establish trends or look for significant differences as experimental parameters are varied. Most data systems allow addition or subtraction of spectra, although division of spectra is also a useful facility to have available. Subtraction of spectra to create so-called difference spectra is widely used in XPS to show up small differences arising from subtle chemical state effects which would otherwise not be noticed.

If difference spectra are to be used, then correct normalization of the spectra is essential if meaningful results are to be obtained. A facility for aligning the energy scales of the two spectra is necessary, together with an appropriate procedure for ensuring that the alignment is performed correctly. Small misalignments on the energy axis can result in large apparent differences on the intensity axis.

As an alternative to subtracting reference data, it can be useful to build model spectra using previously measured data from the pure component materials. This has been found effective for alloys of the transition metals in AES by Langeron (1989). It does involve the approximation that spectra are linearly additive, and care must be taken in cases where strong matrix effects may be expected.

Although not normally available, a facility for division of spectra is essential if recent work on the energy-dependence of instrument transmission and detector efficiencies in AES and XPS is to be exploited to the full (see Chapter 5). Essentially, the requirement is for a spectrum taken under the normal operating conditions for the instrument to be divisible by a standard reference spectrum. The result of the division is the instrument energy-intensity response function. This can then be smoothed and, possibly, parametrized in some way, and subsequently divided into measured data to convert the experimental spectrum into a true energy distribution, with no instrumental contributions to the relative peak intensities. This will greatly facilitate quantitative analysis and should soon become available on the data systems from major manufacturers.

4.8 Differentiation

A standard facility found on many data systems is the ability to differentiate and integrate spectra. In modern high spatial resolution AES systems the beam current into the small spot is often so low, compared with large-spot systems, that conventional analogue modulation methods can not be used and the direct spectrum is detected using pulse counting electronics as discussed in Chapter 3. However, most data banks for AES give spectra in the derivative form. If these are to be used, computer differentiation of the

raw n(E) data is necessary. The differentiation is usually performed in
conjunction with a smoothing function in order to simulate the effect of the
modulation voltage in the older spectrometers. Differentiation, whether by
analogue or numerical means, has the advantage that the small peaks on large
backgrounds, often encountered in AES, are rendered more visible by the
process.

Computer differentiation can give precisely equivalent results to that
expected for modulation methods, provided due care is taken to allow for
differences in the shape of the differentiating function and the modulation
waveform. For example, a 10 V sinusoidal modulation would need to be
simulated by an 11.7 V Savitzky-Golay combined quadratic smooth/differential
function (Seah et al, 1983). A further use of the differentiation facility
in both AES and XPS is to establish the precise energy position of a peak
since a zero crossing point may usually be more accurately measured than the
position of a peak maximum.

4.9 Factor Analysis

Factor analysis is a general and powerful procedure for handling the
large arrays of data commonly found in methods of analytical chemistry such
as infrared spectroscopy. It has recently been recognised to be of use as a
data evaluation method in electron spectroscopy, primarily through the early
work of Gaarenstroom in this area (Gaarenstroom, 1979; 1981; 1982). Details
of the theory are given in the textbook by Malinowski and Howery (1980) and
simplified explanations are given in a number of reviews (Solomon, 1987;
Sherwood, 1990). The method relies for its power on its ability to pick out
common factors - the principle components - in a series of spectra, and to
identify which of these are present in any one spectrum of the set. The
underlying assumption is that the spectra in the set to be analyzed consist
of linear superpositions of standard spectra from known components. Following subtraction of a baseline and renormalisation, the factor analysis
routine works on the whole data set and attempts to determine the number of
components and their respective fractions which can be linearly combined to
represent the total sequence of spectra.

In electron spectroscopy, the method is well suited to cases where
series of similar spectra occur. This may typically be in the analysis of
AES depth profile data, angle-dependent XPS data, or high resolution XPS data
from a sequence of related but differing compounds or polymers showing
chemical shifts. Among the various examples in the literature of the
application of factor analysis to electron spectroscopy data sets, Hofmann
and Steffen (1989) demonstrate its use to gain insights into the structure of
oxide layers on nickel and a nickel based alloy using AES depth profiling,
and Gaarenstroom (1986) illustrates the improvement in the detection limits
in AES depth profiling resulting from a factor analysis approach. Fay et al
(1991) show the use of an improved method of discriminating the number of
factors present using XPS spectra from cobalt aluminate and cobalt oxide
mixtures

Factor analysis is an important technique that is unfortunately not
usually implemented on commercially-available surface analysis data systems.
Nevertheless there are exceptions and it is becoming increasingly recognised
as a valuable practical aid to the interpretation of sets of surface analysis
data.

4.10 Image Processing and Multi-Spectral Acquisition

The discussion of data processing so far has concentrated on the
extraction of information from spectra. However, with modern instruments the

operator is frequently concerned with the information content of an image. In surface analysis by electron spectroscopy, the image will be either an AES or XPS map where the brightness of each pixel represents the intensity of the signal corresponding to a particular element or chemical state on the sample surface at that point. If colour is available, then several elements can be represented at once, or alternatively the concentration of a particular element given in terms of a colour-scale, possibly geographic (as found in an atlas) or thermographic.

The presentation of data in image form opens up a whole new area of image processing for surface analysis. For example, it may be necessary to apply any of the spectral processing methods described in this chapter to each pixel in the image. Such a large amount of processing power would be required that this is never contemplated. Instead, for AES maps, simple peak height measurements are used, although this can be misleading unless steps are taken to remove purely topographical contrast from the images, probably by using the (P-B)/(P+B) method described previously. In XPS images it may be adequate in the case of an intense peak on a weak background to tune the analyser to the peak energy of interest and use the raw signal to form the image, although background subtraction would substantially aid the visibility of the image and is essential for weaker peaks.

There are various types of image processing routines currently available for use in surface analysis. The simplest (conceptually, at least) are those concerned with the way the data contained in the image are presented to the user. Operations including grey-scale manipulations such as histogram equalization and the use of false-colour fall into this category, as do the artificially generated 3-dimensional views. More sophisticated facilities include the 2-dimensional analogues of the various processing routines discussed for spectra above. For example, a two-dimensional low pass filter may be applied to remove high frequency (noise) components in the image, or a high pass filter used as part of a shading correction routine. More complex filtering operations might include optimal, Gaussian, median or so-called sharpening filters, although these are not yet generally available on data-systems for surface analysis. A useful and commonly used image processing routine is the two-dimensional differential which gives enhanced contrast at the edges of features in the image under inspection. Possibly the most sophisticated and general purpose image processing routine is the Maximum Entropy method of image enhancement (Gull and Skilling, 1984). This is extremely powerful in cases which require the deconvolution of a point spread function from a distorted and noisy image in order to reconstruct the data. The key to the method is its rigorous Bayesian statistical foundation and its ability to use all of the available information (not just that contained in the image data-set) in the definition of the point-spread function. The method has already been used in extracting information from noisy spectra, and in the deconvolution of depth profiles from angle-dependent XPS data (Smith and Livesey, 1992); it can not be long before we seen its application to noisy images in surface science.

A novel approach to the acquisition and use of the images formed in scanning Auger microscopy has recently been evolved by Prutton, El-Gomati and co-workers (El-Gomati et al, 1987; Prutton et al, 1991) and by Browning (1985). Known as multi-spectral Auger mapping (MULSAM), it involves the acquisition of multiple images in parallel, using all the signals available from the spectrometer. These may include the Auger peak area itself, the background signal, the sample absorbed current, the fluorescent X-ray signal, the elastically scattered electron signal, and the backscattered electron signal, among other possibilities. Bivariate correlation diagrams are then constructed from the image data-sets. By placing software windows round regions in the correlation diagrams and reconstructing images using only those pixels from the sample producing information which falls within the

window selected, novel insights into the sample structure, and into the
underlying physics of the image formation process, can be obtained.
Alternatively, operations such as topography correction using the
backscattered electron signal may be applied to quantitative Auger maps. The
key to these techniques lies in the parallel acquisition, which gives exact
pixel-to-pixel registration between images derived from the diffferent
signals.

As an example of multi-spectral imaging, El-Gomati et al (1987) measured
Auger maps of the surface of a sample consisting of an aluminium overlayer on
a silicon substrate. They then analysed the data from each pixel in the
image and plotted each point on a scatter diagram with axes corresponding to
the intensity of the aluminium and silicon Auger signals at that point in the
image. Generally, they found pixels to have either low silicon and high
aluminium intensities (electron beam on the aluminium overlayer) or high
silicon and low aluminium intensities (electron beam on the exposed silicon
substrate). However, small clusters of low silicon and low aluminium
intensities and of high silicon and high aluminium intensities appeared in
the correlation diagram. By using only these clusters to reconstruct the
image these areas could be associated with shadowing or enhancement at the
edges of the aluminium overlayer, as shown in Figure 3.10. Other applications include the identification of surface phases, and the separation of
topographical and chemical information. The method is clearly very powerful,
but requires specialised equipment not likely to become incorporated into
commercially available surface analysis instrumentation until some time in
the future.

CHAPTER 5

QUANTIFICATION OF DATA FROM HOMOGENEOUS MATERIALS

5.1 The General Approach to Quantification

Quantification of data from XPS or AES is a type of data processing and could have been included in the previous chapter. However, it can be a complex task and is of such central importance in surface analysis that a separate chapter is required for its discussion. The subject is reviewed in depth by Seah (1990). Currently, there is no one single satisfactory method of quantification which gives reliable results in all cases. Nevertheless considerable progress can be made using a combination of the theoretical results and experimental data that are presented below.

Clearly, it is the ultimate objective of surface analysis to give a quantitative description of the composition of the surface region of the sample under investigation. For this to be achieved, spectral intensities must be related to the number of atoms in the sample emitting electrons which contribute to the spectrum. It is possible to write down equations for XPS and AES which give the emitted intensities in terms of the incident flux, the number of contributing atoms, the cross sections involved, the instrumental and geometrical terms, and other appropriate factors depending on the fundamental physics of the particular processes occuring. The various terms involved are treated in sections 5.4 and 5.5 below. At present, these terms are not known with sufficient accuracy for the first-principles method to be of practical use to the analyst. Instead, for both XPS and AES, it is currently routine practice in surface analytical laboratories to compare intensities in the spectrum from the unknown with reference intensities either obtained by calculation or by measurements of standard spectra of the elements.

Consider the case of a homogeneous binary alloy AB giving two spectral lines with intensities I_A and I_B, in either AES or XPS. The analyst measures I_A and I_B then looks up either in handbooks (3,6,7,8,9) or elsewhere, the intensities I^O_A and I^O_B expected for emission from the pure elements AB and, assuming a linear relationship between atomic fractional composition and signal intensity, obtains the atom fraction of eg element A by comparing relative intensities:

$$X_A = \frac{I_A/I_A^{\,O}}{(I_A/I_A^{\,O} + I_B/I_B^{\,O})} \qquad\qquad [5.1]$$

Generally the I^O values are normalised in some way and are referred to as sensitivity factors S. The method is then usually generalised to:

$$X_i = \frac{I_i/S_i}{\sum_{j=1}^{n}(I_j/S_j)} \qquad\qquad [5.2]$$

This methodology contains a number of approximations and can easily lead to erroneous results. However, it remains current practice in many laboratories. It may be appropriate in cases where routine analysis using the same experimental conditions is being carried out on large numbers of similar samples, and where sensitivity factors have been measured under these conditions using samples of known composition as close as possible to that of the unknown.

For the majority of analysts, it is simply not feasible to compile sets of in-house sensitivity factors suitable for use with the wide range of samples and experimental conditions that will be met in practice. In this case, a set of general routines operating upon a standard data bank of some kind must be employed. For both XPS and AES, the above expressions can be changed to give more accurate results by the inclusion of terms describing the modifications to relative intensities that occur due to effects originating in either the sample or the measurement system. In XPS these terms can be expressed simply as a modification of the sensitivity factor values, whereas for AES additional terms known as matrix factors are required.

Although XPS and AES are essentially surface spectroscopies, it is possible to extract information from below the surface using electrons which have been inelastically scattered, losing energy as they travel from their point of excitation to the surface. Wertheim (1990) has pointed out that the area of the entire photoemission line, including the loss tail, is proportional to the photoelectron cross-section. Furthermore, the overall shape of the loss structure is governed primarily by the bulk composition of the specimen, and is largely insensitive to surface segregation and contamination. Therefore, argues Wertheim, it is possible to use single measures of loss intensities at suitable energies below primary peaks which, when corrected for the (known) photoelectron cross-sections, give directly the bulk composition of the sample. This is an attractive proposition, however, it does appear to be somewhat at odds with the work of Tougaard (1989) in which it is demonstrated that the details of the loss structure may be used to give information on both the amount of material present and its distribution in depth. This use of XPS and AES for the non-destructive determination of composition against depth profiles is discussed in Chapter 6.

5.2 Statistical Treatment of Cumulative Errors

It is standard practice in many of the "bulk" methods of analytical chemistry to give not only the composition as determined by that technique, but some estimate of the confidence that can be associated with it. The method of quantitative analysis outlined above is subject to systematic and random errors. However, these are almost invariably ignored and no attempt made to give the confidence limits on a particular measurement. In electron

spectroscopy, systematic errors arise largly from errors in the numerical
values of the relative sensitivity factors used, while random errors originate from noise in the data. Errors of measurement in AES and XPS are
discussed by Powell and Seah (1990). The treatment of statistical uncertainties in the data and of the uncertainty in the relative sensitivity factors
has been the subject of recent research (Harrison and Hazell, 1992; Cumpson
and Seah, 1992 (two papers); Evans, 1992) and is discussed here in order to
show how confidence limits may be estimated for quantitative surface
analyses. The separate question of what are the best sets of sensitivity
factors to use is covered in sections 5.4 (for XPS) and 5.5 (for AES) below.

In surface analysis by electron spectroscopy, the uncertainty of the
final result of quantification is a result of both systematic and random
factors. Sources of systematic error include the assumption of sample
homogeniety, incorrect measurement of peak intensities by using an
inappropriate background subtraction routine, inadequate correction for
instrumental terms, and inaccuracies in the relative sensitivity factors
themselves. In many cases, particularly for peaks with good counting
statistics, the systematic errors dominate, and random errors due to poor
counting statistics are not significant. Nevertheless there are many
instances where a knowledge of the random error is important. An example may
be where compositional trends between similar samples are studied and it is
necessary to know the level of apparent compositional difference that is of
real significance. Similarly, where elements are embedded at low concentration in a matrix, the problem of the detection limit is encountered and some
estimate of a meaningful signal level is required. The systematic errors
mentioned above are discussed in the relevant sections elsewhere in this
book. This section concentrates on the statistical contribution to
uncertainty arising from noise in the data.

Data are always subject to noise. For electron spectrometers employing
pulse counting signal detection, the principle source of uncertainty is the
statistical variation in the measured count rate. Ideally, the count rate
distribution should follow a Poissonian form, however, for reasonable count

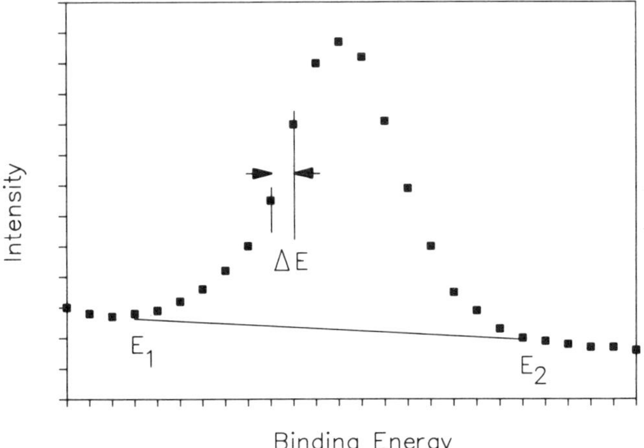

Figure 5.1 Schematic of a peak in electron spectroscopy, showing the various
parameters required for the determination of the error in a
quantitative analysis due to random noise

rates this approximates rather well to a Guassian distribution, and in this
discussion Gaussian distributions will be assumed. It is worth checking that
the count rate from an individual spectrometer does follow this distribution.
If the discriminator setting and multiplier voltage are away from optimum in
a channel electron multiplier system, significantly different distributions
can be obtained which will invalidate any serious attempt to place statistical confidence limits on the results of a quantitative analysis.

Harrison and Hazell (1992) have examined the case of the random error in
either a peak height or a peak area measurement in an experiment obeying
Gaussian statistics. For the generalised peak shown in Figure 5.1, they show
that the intensity I to be used in the quantification equation [5.2] above,
corresponding to the measured peak area A, is given by:

$$A = (\Delta E/t)[\sum_{i=2}^{n-1} N_i - (n - 2)(N_1 + N_n)/2]^{\frac{1}{2}} \tag{5.3}$$

where ΔE is the energy width for a channel over which data are recorded for
time t, N_1 is the background count rate at energy E_1 at one side of the peak
and N_n is the corresponding background at the other side of the peak for a
peak containing n channels. Noting that the peak area consists of the
background area subtracted from the total signal contained under the peak
envelope, they then apply the usual statistical rules for the combination of
errors to show that the random error associated with the peak area
measurement, $\delta(A)$ is given by:

$$\delta(A) = (\Delta E/t)[\sum_{i=2}^{n-1} N_i + (n - 2)^2(N_1 + N_n)/4]^{\frac{1}{2}} \tag{5.4}$$

Using the relative precision $\delta(A)/A$ and applying partial differentials to
equation [5.2] allows the random error in the quantification to be estimated
as:

$$\delta_R(X_i) = [(1 - 2X_i)^2\,_R^2(A_i) + \sum_{j \neq i}X_j^2\,_R^2(A_j)]^{\frac{1}{2}} \tag{5.5}$$

Similar equations are derived by Cumpson and Seah (1992) and by Evans (1992).

An XPS spectrum typical in overall form of many that may be met in
practice is shown in Figure 5.2. The spectrum shows carbon and oxygen at
relatively high intensities, together with significant amounts of aluminium,
sulphur and magnesium. Nitrogen is also present, and a trace of copper is
detected. The signal and background levels, together with the sensitivity
factors used in the quantification routine (Equation [5.2]) are shown in the
Table 5.1 to illustrate the results of the statistical analysis described
above. The precise values of the sensitivity factors used are not important
here; they are discussed further in subsequent sections of this chapter. The
final column in Table 5.1 shows the overall precision determined from
Equation [5.5] above, expressed in atomic percentage.

In Table 5.1, the precision of the quantification is given in atomic
percentage, rather than relative precision, in order to bring out the
significance of the analysis. The precision of the oxygen and carbon
determinations appear to be similar at around 0.6 atomic percent. However,
the relative precision on the carbon is rather poorer, corresponding to a
relative error of approximately 2% as opposed to one of around 1% on the
oxygen. The errors in the copper and nitrogen determinations correspond to
significantly large fractions of the amounts present. This is not
surprising, particularly for the copper, where a very small peak is found in
the binding energy region of the spectrum above the oxygen Auger peak, where

Table 5.1 Quantitative analysis of the XPS spectrum shown in Figure 5.2.   The compositions are obtained from the peak areas and relative sensitivity factors using Equation 5.2, and the uncertainties are estimated using the method of Harrison and Hazell (1992).

Element	Area	Back-ground	RSF (S)	X (at%)	σ(X) (at%)
Cu	2220	22908	11.898	0.21	0.07
O	98290	7900	2.641	42.75	0.55
N	1670	4922	1.712	1.12	0.25
C	26510	4610	1.000	30.45	0.57
S	8620	3384	1.794	5.52	0.24
Al	7680	1792	0.604	14.61	0.52
Mg	1770	1620	0.381	5.33	0.73

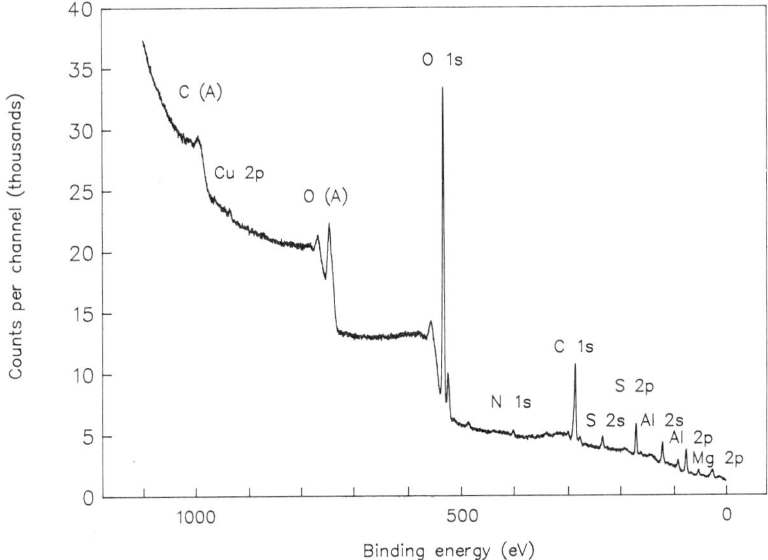

Figure 5.2 An XPS spectrum of a multi-component sample based on aluminium oxide, acquired using magnesium radiation.   The peak areas, quantitative composition and the associated errors calculated using the method of Harrison and Hazell (1992) are shown in Table 5.1

a very high background is present. Had the spectrum been recorded with aluminium radiation, the copper 2p photoelectron peak would have appeared at a lower binding energy than that apparent for the oxygen Auger peak, resulting in the same signal intensity (after correction for the difference between sensitivity factors for aluminium and magnesium radiations) but a lower background and, correspondingly, a higher precision. For elements at these low concentrations where the error can approach the magnitude of the measurement, it may be more appropriate simply to record them as detected, rather than attempt a quantification.

Harrison and Hazell (1992) provide illustrations of the implications of their analysis on the kind of stategy that may be adopted for a particular experiment. An, at first sight, surprising result is that if the amount of a particular element in the sample is to be determined to high accuracy then, because of the complex way in which the errors interact, it is necessary to acquire good quality data not only on the element of interest but also on the others present.

The discussion above has assumed that the peaks in the spectrum which is to be quantified contain only one contribution. In practice, it is often desirable to separate out the various contributions to a peak from an element present in the sample in a number of chemical states. In this case, the accuracy of the peak fitting can become a major source of random error in the peak areas that are extracted and input to the quantification routine.

The statistical uncertainties involved in peak fitting are discussed by Evans (1992) and Cumpson and Seah (1992). Cumpson and Seah (1992) concentrate on providing a definitive statement of the principles involved in the estimation of uncertainties in curve fitting and show how standard commercially available software can be used to provide these estimates, whereas Evans (1992) describes a rapid but approximate method of achieving this end using a minimal amount of computation.

It is expected that manufacturers of commercial data systems for AES and XPS will soon incorporate automatic routines for estimating the precision of measurements made in their spectrometers. At the moment, it appears very straightforward to implement the equations of Harrison and Hazell (1992) and provide estimates of the errors involved in quantification of spectra where curve resolving is not necessary. However, providing similar estimates where features in the spectrum have to be individually resolved by curve fitting may be clear in concept but is possibly quite expensive in terms of computing time. This may inhibit its take up.

5.3 Attenuation Lengths in Electron Spectroscopy

The concept of electron attenuation length was introduced in Chapter 2. It is of such fundamental importance in quantitative electron spectroscopy that it is worth some elaboration here. The attenuation length, $\lambda(E)$, is usually thought to determine the information depth of AES, XPS and other surface electron spectroscopies through the approximate relation

$$I = I_0 \exp (- d / \lambda(E) \cos\theta) \qquad\qquad [5.6]$$

where I/I_0 is the reduction in signal strength at energy E experienced as a result of the generation of electrons at depth d and their subsequent emission at an angle θ relative to the surface normal.

The attenuation length was previously assumed to be synonymous with the inelastic mean free path, defined as the mean distance an electron of energy E could travel in any particular material before loosing energy in an inelastic collision. However, it is now known that elastic collisions are also important in determining the information depth, and that consequently the attenuation lengths are shorter than the inelastic mean free paths by a factor of around 1.15 to 1.30 (Jablonski, 1987). More accurate work has shown that Equation [5.6] above is only an approximation to the attenuation of signal as a function of depth, and a more precise description requires the use of a complex depth distribution function, whose form may be a function of electron emission angle as well as material and energy. This is discussed further below.

In the context of Equation [5.6], Seah and Dench (1979) have published a compilation of measured attenuation length data, referred to as inelastic mean free paths, to which universal curves were fitted in order to give a means of estimating unknown attenuation lengths. They give:

for elements –

$$\lambda(E) = 538aE^{-2} + 0.41a^{3/2}E^{1/2} \quad nm \qquad [5.7]$$

for inorganic compounds –

$$\lambda(E) = 2170aE^{-2} + 0.72a^{3/2}E^{1/2} \quad nm \qquad [5.8]$$

and for organic compounds –

$$\lambda(E) = (10^3/\rho)(\; 49E^{-2} + 0.11E^{1/2}\;) \quad nm \qquad [5.9]$$

where ρ is the bulk density in gcm^{-3}, and a is the average monolayer thickness in nm given by:

$$a^3 = 10^{27}Z\;/\rho\,nN_A \qquad [5.10]$$

where Z is the atomic number, n is the number of atoms in the molecule and N_A is Avogadro's Number.

Equations [5.7], [5.8] and [5.9] give a fit of $\lambda(E)$ to the measured data to within a standard deviation of around 35%. Most of the data included in the compilation by Seah and Dench (1979) were obtained by measuring the attenuation of electrons transmitted through thin overlayers as a function of the thickness of the overlayer. The overlayer thicknesses were typically estimated using a quartz crystal microbalance and hence are average thicknesses. Thin films are known to grow with a variety of structures, often involving the production of islands of some sort. Island structures lead to uneven films, therefore the data analyzed by Seah and Dench (1979) contain uncalibrated systematic errors. Further, more recent analyses have tended to give energy dependencies of the attenuation lengths rather stronger than the $E^{0.5}$ of equations [5.8] to [5.10]. Wagner et al (1980) find an energy dependence ranging from $E^{0.54}$ for gold to $E^{0.81}$ for silicon. Energy exponents in the range 0.65 to 0.75 are typically found by calculation (Ashley and Tung, 1982; Payling and Szajman, 1987; Powell, 1988). Using Monte-Carlo calculations, Ebel et al (1988) give a mean calculated exponent of 0.80 +/- 0.04 for attenuation lengths of 28 elements, and 0.76 +/- 0.04 for inelastic mean free paths for the same set of elements. Thus, not only are the attenuation lengths and inelastic mean free paths different, but their energy dependencies also differ.

Rather than providing a best fit to experimental data, more recent work has focused on calculating the inelastic mean free path from first principles. Tanuma, Powell and Penn (1987, 1988) use a model dielectric function fitted to experimental optical constants to give an inelastic scattering probability as a function of energy loss. They then employ the Bethe equation (Bethe, 1930) to give the energy dependence of the calculated inelastic mean free paths. The constants in the Bethe equation are fitted empirically to materials parameters to give the following relation for the inelastic mean free path:

$$IMFP = E \; / \; E_p^{\,2}\beta\ln(\gamma E) \quad \text{Å} \qquad [5.11]$$

Here, E_p is the free electron plasmon energy given by:

$$E_p = 28.8(N_v\;/Z)^{1/2} \qquad [5.12]$$

where N_v is the number of valence electrons per atom or molecule, and β is given by:

$$\beta = -2.52\text{x}10^{-2} + 1.05 \ (E_p^2 + E_g^2) + 8.10\text{x}10^{-4}\rho \qquad [5.13]$$

in which E_g is the band-gap energy in eV, and γ is given by

$$\gamma = 0.151\rho^{-0.49} \qquad [5.14]$$

Over the energy range 200 - 2000 eV this gives an accuracy of around 12%, but in practice involves the uncertainty between that which is calculated (the IMFP) and that which, in the context of Equation [5.6], is important in most experimental situations (the attenuation length). This work has been extended to include the energy range down to 50 eV (Tanuma et al, 1990; 1991)

Work by Jablonski and Ebel (1988), Werner (1991), Werner et al (1991) and Jablonski and Tougaard (1990) has shown that the concept of a single emission angle independent attenuation length is flawed, particularly at higher emission angles, because of the presence of elastic scattering. Werner and Stori (1992), using an empirical fit to the results of Monte-Carlo calculations, propose that the depth distribution function for AES and XPS be defined by:

$$\phi \ (z,\theta) = - \ a_1\exp \ (-z/\lambda_t \cos \theta) + a_2(\theta)\exp \ (-z/\lambda_a\cos \theta)$$
$$+ \ a_3(\theta)\exp \ (-z/\lambda_a) \qquad [5.15]$$

where λ_a is known as the attenuation parameter and corresponds to the slope of the depth distribution function at great depths (on a logarithmic scale), and a_1, a_2 and a_3 are normalisation constants given by:

$$a_1 = (\lambda_t/\lambda_i)(\lambda_i - \lambda_a)/(\lambda_a - \lambda_t) \qquad [5.16]$$

$$a_2 = 1 + a_1 - a_3(\theta) \qquad [5.17]$$

$$a_3 = (\lambda_t/\lambda_i)(\lambda_i - \lambda_a) \cos \theta \ /\lambda_a(1 - \cos \theta) \qquad [5.18]$$

Here, λ_i is the conventional inelastic mean free path and λ_t is the total mean free path. Calculations of these parameters are given by Werner and Tilinin (1992).

Werner et al (1991) point out that as a consequence of elastic scattering, the shape of the depth distribution function is a function of depth, as described by the equations above. In the near surface region they find a depth distribution function with an attenuation length corresponding to the inelastic mean free path. At intermediate depths between one and three times the inelastic mean free path, the effective attenuation length changes from the inelastic mean free path to the attenuation parameter λ_a. Below this, the electron trajectories are fully randomised and the slope of the depth distribution function corresponds to an emission angle independent λ_a.

The implications of this work for practical quantitative surface analysis are not yet fully worked out, and there has been some criticism of the work of Werner et al (Dwyer and Richards, 1992). Nevertheless, it is reasonable to expect any new definition of escape depth in electron spectroscopy to incorporate the effects of elastic as well as inelastic scattering of the outgoing electrons. For the moment, it seems as if the calculations of Tanuma et al (1991, 1991), with interpolation where appropriate, should be used, but with due allowance made for the reduction of

the escape depth due to elastic scattering. However, recent work has shown
that, for both AES and XPS, the inelastic mean free path should be used in
preference to the attenuation length when quantifying spectra using simple
methods based on intensity ratios (Jablonski, 1990; Jablonski and Powell,
1993). It is worth mentioning that the figure for carbon given by Tanuma et
al (1991) is calculated using the density of amorphous carbon and that for
hydrocarbon overlayers the true figure could be significantly higher.
Determinations of attenuation lengths in hydrocarbon films are given by Bain
and Whitesides (1989). Their energy dependence is discussed by Ebel et al
(1991), who find an average energy exponent of 0.64.

5.4 Accurate Quantification in XPS

The method of quantification outlined in section 5.1 is routinely used
in surface analysis laboratories, despite being oversimplified and prone to
error. In this section, the various contributions to intensities measured in
XPS are discussed in order to arrive at a systematic and consistent method of
quantifying XPS spectra from homogeneous samples of unknown composition.

5.4.1 Calculations of Intensity

The intensity of a particular photoelectron line in an XPS spectrum is
determined by the flux of incident X-rays, the cross-section for photoemission of the particular level involved, the probability of that photoelectron escaping from the sample, and the number density of atoms contributing to the line, together with terms describing the efficiency of detection
of the photoelectron and the geometry of the experiment. The intensity
expected in a photoelectron line at kinetic energy E_A from a pure reference
sample of element A is given by:

$$I^O_A = \sigma_A(hv)\ J_o\ L_A(\gamma)\ Q(E_A)\ N^O_A \lambda_A(E_A)\cos\theta \qquad [5.19]$$

where $\sigma_A(hv)$ is the cross section for photoemission for the line of interest
from element A at photon energy hv, J_o is the X-ray photon flux at the sample
surface, $L_A(\gamma)$ is the angular asymmetry parameter for the photoelectron line
concerned and an included angle γ between the incoming photons and the outgoing photoelectrons, $Q(E_A)$ is a term describing the transfer characteristics
of the spectrometer, N^O_A is the number density of atoms in the pure reference
sample, $\lambda_A(E_A)$ is the attenuation length for photoelectrons of kinetic energy
E_A in element A, and θ is the angle of emission of the photoelectrons
measured from the sample normal. The X-ray flux is assumed to be constant
across the sample surface, and the analyzer is assumed to collect at one
angle only. If these assumptions are not valid then the appropriate
integrals must be included in Equation [5.19]

Photoelectron cross sections, normalised to the carbon 1s line, have
been calculated by Scofield (1976) for magnesiun and aluminium Kα radiations,
and are believed to be accurate to around 5%. Values for peaks commonly
observed in XPS are shown in Figure 5.3, calculated for aluminium radiation.

The angular asymmetry parameter depends upon the angle between the X-
rays and the electron detection direction, and is given by

$$L_A(\gamma) = 1 - \beta_A(3\cos^2\gamma - 1)/4 \qquad [5.20]$$

where β_A is a constant which has been calculated for photoelectron peaks of
importance in XPS by Reilman et al (1976). Clearly, $L_A(\gamma)$ becomes unity when
an angle is used such that $\sin\gamma = 2/3$. For this reason, some spectrometers
are constructed at or near this angle so that the angular

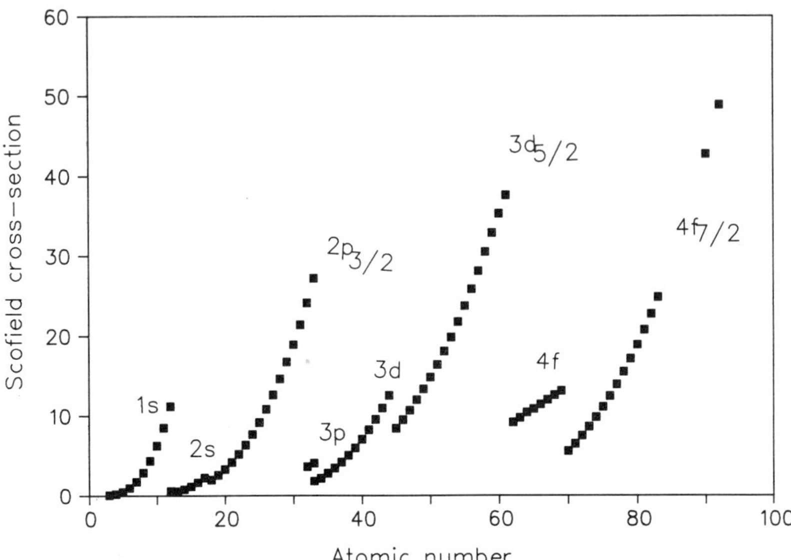

Figure 5.3 Cross-sections for the principle photoelectron lines used in XPS,
calculated for Al Kα radiation (Scofield, 1976)

asymmetry parameter need not be included in the expressions used for
quantification. However, recent calculations have shown that the apparent
value of β_A is reduced by elastic scattering to an effective value β_A^* which
depends upon the atomic number Z according to (Jablonski, 1989)

$$\beta_A^* = (0.781 - 5.14 \times 10^{-2}Z + 3.1 \times 10^{-5}Z^2)\beta_A \qquad [5.21]$$

These effects are illustrated for a typical included angle of 70° in
Figure 5.4, which shows the calculated values of $L_A(\gamma)$ with and without the
inclusion of elastic scattering. Unless the spectrometer geometry is far
from the 'magic angle' of 54.7° the effect of $L_A^*(\gamma)$ is small and is usually
neglected.

Until recently, the analyzer transmission and detection efficiency term,
Q(E), has been rather difficult to deal with. Most practitioners have
assumed energy dependencies of the form $E^{-0.5}$ or E^{-1}, relying upon the
results of calculations for ideal instruments (Seah, 1980). In other work,
the energy dependence of this term has simply been ignored. However,
reference spectra for XPS which do not contain any contribution due to the
energy dependence of the instrument function have now been published (Seah
and Smith, 1990). The energy dependence of Q(E) for any XPS instrument may
be derived by measuring spectra for copper, silver and gold using magnesium
or aluminium radiation under the normal operating conditions for that
instrument and dividing by the appropriate reference spectrum. The method
has been tested and verified using interlaboratory comparisons, and gives a
precision of within a few percent. It is discussed in Chapter 3, Section
3.3.4 above.   It is important to realise that the Q(E) so derived is a
sensitive function of the operating condition of the instrument and will
change between low and high resolution settings, or between large and small
area analysis (Smith and Seah, 1990).

The number density of atoms in the sample is the reciprocal of the
atomic volume, a_A, and may be easily calculated from the density, atomic mass
and Avogadro's Number.

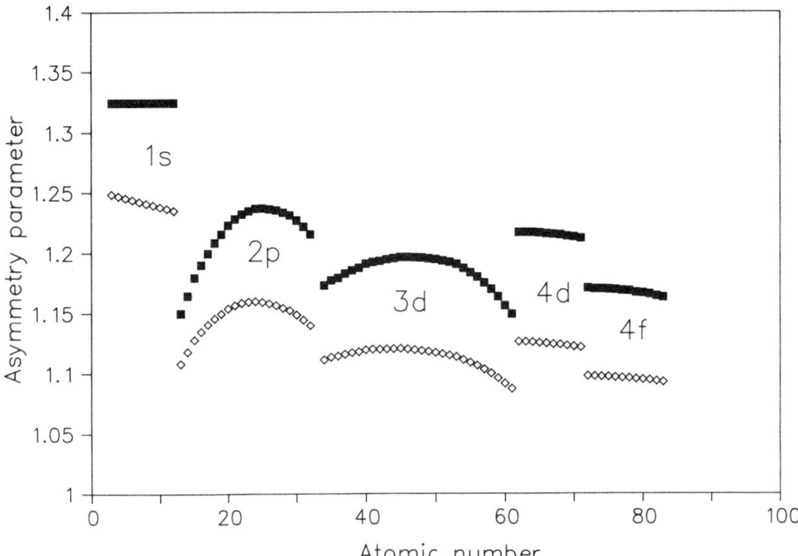

Figure 5.4 Calculated values of the angular asymmetry parameter. The filled symbols are the values for a spectrometer with an included angle between the X-ray source and the analyzer of 70°, and the open symbols show the effect of correction for elastic electron scattering using equation [5.21] (Jablonski, 1989)

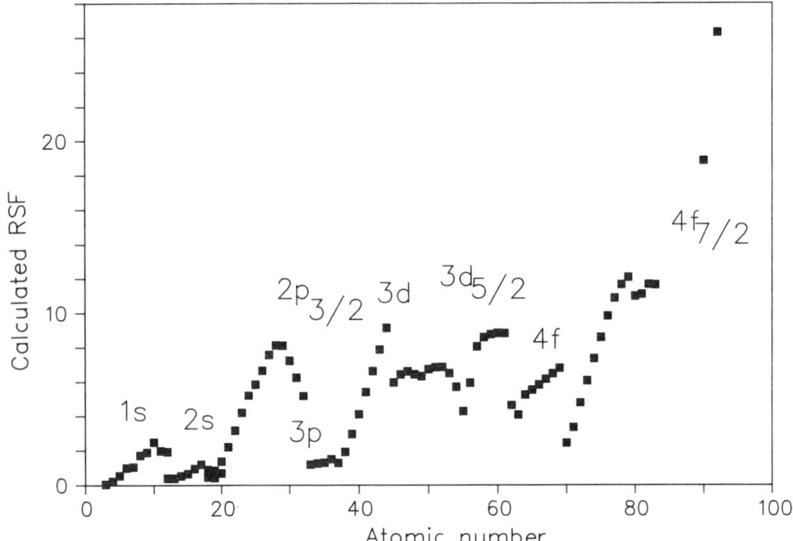

Figure 5.5 Calculated relative sensitivity factors for the principle photoelectron lines used in XPS, obtained using the method discussed in the text

The attenuation length $\lambda_A(E_A)$ for photoelectrons at energy E_A in the pure element A is obtained from the result of calculations discussed in Section 5.3 of this chapter.

5.4.2 Accurate Relative Sensitivity Factors

Section 5.1 above introduced the general method of quantification of electron spectroscopic data using relative sensitivity factors. It is clear that, with the knowledge above, equation [5.19] may be used to calculate relative sensitivity factors for a particular instrument operating condition and geometry. As the relative sensitivity factors are always used in a ratio such as equation [5.2] above, the X-ray flux and $\cos \theta$ terms in equation [5.19], which are constant for a particular experiment, cancel out, and the relative sensitivity factor for a given photoelectron line for element A can be written

$$S = \sigma_A \, L^*_A(\gamma) \, Q(E) \, \lambda_A(E_A) \, a_A^{-3} \qquad\qquad [5.22]$$

Sensitivity factors calculated using equation [5.22] for an analyzer with an assumed energy dependence of $E^{-0.5}$ are shown in Figure 5.5, normalised to the carbon 1s line.

Despite the simplicity of the derivation of S using this approach, there is no universal agreement on the best choice of relative sensitivity factors for a particular experimental case. Many different sets of S values are available in the literature, with significant differences between them. Experimental measurements are reported by Wagner (1972, 1977), Nefedov et al (1973, 1975), Adams et al (1977), Evans et al (1977, 1978), and Ward and Wood (1992), among others. The handbook data given by Wagner et al (1979) is derived from a simplified form of equation [5.22] above, whereas Wagner et al (1981) present an empirical data set based upon measured values of intensity fitted to smooth curves obtained from equations like [5.22]. A detailed comparison of the earlier data sets is given by Seah (1980) and reviewed by Seah (1986).

The relative accuracy of the different methods of obtaining relative sensitivity factor datasets has been investigated by Battistoni et al (1985). They found that accuracies around 10% could be achieved using relative sensitivity factors calculated from first principles (as above) or measured from reference samples in the same instrument. However, a much lower accuracy, as expected, was obtained when relative sensitivity factors reported in the literature were used directly. This was attributed primarily to differences in the response functions of the different instruments used to produce the data.

The greatest single cause of discrepancy between measured and calculated relative sensitivity factors, assuming instrument response functions are known and properly corrected for, is that, often, the measurement does not include all of the photoemitted intensity. Shake-up and shake-off effects can result in the intensity of a peak being distributed amongst satellites extending over a wide energy range. Measured intensities are usually determined by a simple background subtraction process over the most intense component of the peak, which can result in the loss of intensity if the satellites are not properly included. This effect has been investigated by Halbritter et al (1988), who demonstrate that, in extreme cases, up to 80% of the intensity may be lost. Furthermore, examination of the Tougaard background (discussed in Chapter 6) for metallic elements shows that, in some cases, a simple straight line or Shirley background may only account for around 30% of the true peak intensity.

In view of the uncertainties involved, the question of which is the best set of relative sensitivity factors to use for a particular analysis is not easy to answer. However, Seah (1986) concludes, from a detailed analysis of the available data, that the hybrid approach homogenising theory and experiment used by Wagner et al (1981) represents good common sense. This dataset has the advantage of being the largest available, and is reproduced in the book by Briggs and Seah (1990). It should be noted that an instrument transmission function varying as the inverse of the electron kinetic energy is used, and a suitable correction should be applied when the instrument response function is known to follow a different behaviour.

An interesting variation on the method described above is the multiline approach of Hanke et al (1986). They note that, to get maximum information from the XPS spectrum, all lines present should be included in the quantification routine. The measured spectral intensities are weighted according to their relative strengths, and the composition determined by solving a system of linear equations in such a way as to minimise the discrepancies between compositions calculated using the (weighted) individual lines. Despite its advantages, the method appears complex and does not seem to have been widely adopted.

5.4.3 The Measurement of Intensity

So far in this discussion of quantification, no mention has been made of exactly what are the intensities to measure. This is a non-trivial problem which has not yet been fully resolved. The answer is probably to use a measure that is as close as possible to that which was employed to determine the original I^o values that are to be used. In XPS, the principal peaks are often very sharp, with an apparent height above the background that is therefore rather sensitive to the resolution, scan rate and time constant of the spectrometer. However, the area of a peak should be less sensitive to these instrumental factors, as a decrease in height will be compensated for by an increase in peak width. Therefore, relative sensitivity factors for XPS are always quoted for peak area measurements. This can present a problem in the unlikely event of a computer data processing system not being available; in the past analysts have resorted to counting squares on graph paper, or even cutting out the peaks and weighing them, in order to arrive at some estimate of the peak area.

A rather more common difficulty is that of determining the background to subtract from the peak. There appears as yet to be no single satisfactory method of carrying out this operation. The analyst is normally faced with a choice between a straight line or the methods of Shirley (1972) or Tougaard (1989). Instances exist where each one of these appears best. The details of these methods are discussed in Chapter 4, Section 4.5. For practical purposes it is probably more important that, whatever background subtraction routine is chosen, it is used consistently. Errors will then tend to cancel out when peak area ratios are taken.

5.5 Accurate Quantification in AES

5.5.1 Use of Direct or Derivative Spectra in AES

In the early stages of the development of the AES technique, it was found convenient to incorporate AES into existing low energy electron diffraction (LEED) systems, and to use the LEED optics as a retarding field analyser.

For a description of the LEED technique see Clarke (1985). The signal to noise ratio could be improved by modulating he voltages on the grids of the optics and using a lock-in amplifier for signal detection in such a way that the first derivative of the spectrum was measured (Weber and Peria, 1967). This had the advantage of partially removing the high background from the spectra, therefore rendering the peaks more visible.

As we have seen, the trend in modern AES instruments is towards ever higher spatial resolution, with a concomitant reduction in beam current. Generally, beam currents are at such a level that pulse counting detection must be used in place of the analogue mode used in the earlier instruments. The result is that spectra are commonly acquired in the direct rather than the derivative form. However, over a period of time, spectroscopists have become familiar with the differential mode and such is the strength of tradition that it is normal practice in many laboratories to acquire the spectrum in the direct form and then to differentiate it numerically using a computer. This is despite the fact that both the background and the precise details of the peak shapes, which are lost in the differentiation process, contain much useful information to aid the analyst. For accurate computer differentiation the spectrum must be measured over an energy range of several electron volts either side of the peak, adding to the acquisition time. Differentiation also introduces a further source of uncertainty when quantifying data, although it does have the advantage of effectively removing the background, for which there are as yet no completely satisfactory subtraction methods.

There are occasions where it may be desirable to monitor the peak to peak height in the derivative spectrum. For example, when depth profiling using argon-ion sputtering a simple and rapid measurement of peak intensity is usually all that is required. In this case, for reasons concerning the signal to noise ratio attainable, it is almost always better to increase the electron beam spot size, and hence the beam current, and to modulate the spectrometer thereby obtaining the derivative spectrum directly, rather than using computer differentiation.

When using the differential spectrum, care must be taken to ensure that the measured spectrum is acquired under the same conditions of resolution and modulation (or differentiation) as the reference data otherwise erroneous conclusions can easily be drawn. For example, Figure 5.6 shows two spectra taken from the same contaminated copper surface at high and low resolutions (Anthony and Seah, 1983). The low resolution spectrum shows an apparently much higher degree of contamination, due to a reduction in the peak-to-peak height of the sharper high energy copper peaks in this differentiated spectrum. The relative peak areas in the direct spectrum would, of course, have remained constant. Changes in the modulation voltage or the width of the differentiating function can have equally serious effects. The analyst must ensure that the reference data and the data to be analyzed are part of a consistent set, otherwise the measurement of relative peak-to-peak height in the differential spectrum fails as a basis for quantification.

As was the case for XPS, the important thing as far as quantification is concerned is to define precisely what measure of intensity is to be used, and to use it consistently. If relative sensitivity factors are to be used, then it is important that, as far as is possible, they use the same measure of intensity as is employed in the experimental determination of measured intensity from the unknown. However, in a systematic study of different procedures for the quantitative analysis of mixtures by AES, Turner and Lee

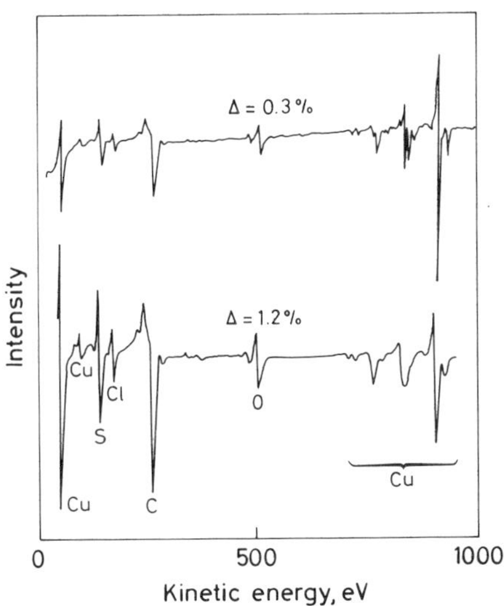

Figure 5.6 Differential Auger spectra acquired from the same contaminated copper surface in the same spectrometer with the same modulation but different resolution settings (Anthony and Seah, 1983). Reproduced with permission, Elsevier Science Publishers B.V., Amsterdam

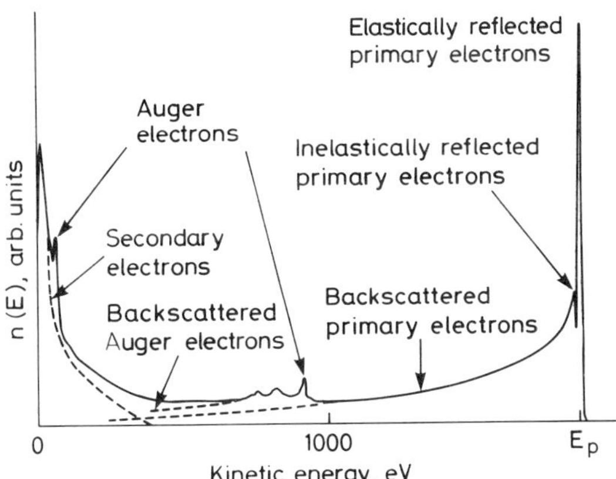

Figure 5.7 Schematic diagram showing the principal contributions to the measured intensity in an Auger electron spectrum.

(1986) found that, when only analogue data were available, there were advantages to using the negative peak-to-background intensity rather than the peak-to-peak intensity.

If the direct spectrum is to be used quantitatively, then a reliable method of background subraction is required. The Auger electron spectrum contains several distinct contributions, as outlined schematically in Figure 5.7. The high energy end of the spectrum shows the elastically reflected electrons from the primary beam, together with their associated loss structure. The intensity in the spectrum then falls away gradually at lower kinetic energies, as a result of rediffused primary electrons that have been backscattered and emitted from the sample after losing energy in inelastic collisions. At lower energies, the intensity in the spectrum increases again, due to the emission of true secondary electrons. Somewhere between these two energy regimes fall the Auger peaks of interest, usually at energies where the background contains contributions from both backscattered and secondary electrons. Each Auger peak also acts as a source of electrons which may be inelastically scattered and therefore contribute to the background in the lower kinetic energy part of the spectrum. The exact nature of these various contributions to the spectral background in AES has been the subject of much intensive research (Tougaard, 1989; Jousset and Langeron, 1987; Ichimura et al, 1988; Browning et al, 1985) and, while the situation is not yet clear, some general conclusions can be drawn.

The background contribution due to backscattered primary electrons, $B(E)$, appears to be quite well described by a simple exponential of the form

$$B(E) \propto \exp(E_p - E) \qquad\qquad [5.23]$$

over the range 0.2 to 0.8 E_p, where E_p is the primary beam energy (Jousset and Langeron, 1987). The secondaries follow a cascade process to give a contribution $S(E)$ such that

$$S(E) \propto E^{-m} \qquad\qquad [5.24]$$

here m is a constant typically around 1.5 (Browning et al, 1985). In principle, the backscattered electrons can be removed by subtracting a suitably fitted exponential and then the spectrum may be linearized by plotting log (intensity) vs log (energy) to give a straight line of slope -m, away from the Auger peaks. This straight line could be be subtracted from the peak region of interest to give a peak area. Unfortunately it is not yet clear how to determine the fit of this straight line on the low energy side of the peak and it is not yet a practical procedure for routine analysis. More complicated procedures which appear to function well are given by Tougaard (1989) and by Rosenberg et al (1988). These methods are likely to be implemented on commercially available data systems in the near future. For practical purposes, the analyst wishing to measure peak areas in AES is probably best advised to follow the same procedure as for XPS. That is, use a straight line or Shirley background with a combination of judgement, experience and intuition to fix the end points.

It should be emphasised that before any of the methods involving linearization of the secondary electron background or correction for the slope due to the rediffused primaries are applied to the experimental data, all contributions from the energy-intensity response function of the spectrometer must be removed. This is a non-trivial procedure, best carried out with the aid of standard reference spectra (Smith and Seah, 1988) as discussed in Chapter 3 Section 3.3.4. The effect of the energy dependence of the analyzer sensitivity on calculated Auger sensitivity factors is discussed by Mroczkowski (1989).

An interesting recent development which avoids the problems of background subtraction is the suggestion that, for AES, linear combinations of direct elemental spectra may be used to simulate spectra of multi-component systems (Langeron, 1989; Ze-jun et al, 1987). The fractional composition of the unknown is then determined from the relative proportions of the elemental spectra required to give a good fit to the data. This proposed method appears to give good results for simple alloying systems but requires careful control of the experimental conditions.

5.5.2 Calculations of Auger Intensities

The intensity emitted into a single Auger peak for unity incident flux depends upon the total ionization cross-section, the probability of Auger emission and the proportion of rediffused primary electrons that still have sufficient energy to cause ionization, as well as the terms alrady discussed for XPS involving the analyzer transmission and detection efficiency, and the depth distribution of the element of interest. Therefore, for unity incident flux, the intensity of an Auger peak at energy E_A due to element A in matrix M with a depth distribution $N_A(z)$ is given by:

$$I_A = \sigma_t(E_p) \, \gamma_A \, \sec \theta_i \, [1 + r_M(E_A, E_p, \theta_i)]$$

$$Q(E) \int_0^\infty N_A(z) \, \exp(-z/\lambda \cos\theta_o) dz \qquad [5.25]$$

where $\sigma_t(E_p)$ is the total ionization cross section for a primary beam of energy E_p, γ_A is the probability of Auger emission for the transition of interest, θ_i is the angle of incidence measured relative to the sample normal, $r_M(E_A, E_p, \theta_i)$ is the backscattering factor for Auger production at energy E_A for a primary beam of energy E_p incident at θ_i, $Q(E)$ is the analyzer transfer function discussed in Chapter 3, and θ_o is the angle of emission of Auger electrons. Assuming a uniform depth distribution and performing the integration over depth gives the simplified form of [5.25]:

$$I_A = \sigma_t(E_p) \, \gamma_A \, \sec\theta_i \, [1 + r_M(E_A, E_p, \theta_i)] \, Q(E) \, N_A \lambda \cos\theta_o \qquad [5.26]$$

Various authors have given expressions for $\sigma_t(E_p)$. For AES, the most relevant works are that of Bethe (1930) and Gryzinsky (1965). In both cases these give curves for the normalised cross section which rise rapidly to a peak at a primary energy around 4 to 5 times the energy of the level to be ionised, falling off with energy rather slowly thereafter. Measurements of cross sections, and their correlation with theory are reviewed by Powell (1989)

The probability γ_A of Auger emission from an ionized level depends upon the binding energy of the particular core level concerned. Non-radiative decay through the Auger process competes with radiative decay (X-ray flourescence), however, for binding energies below around 2 keV, Auger emission dominates. As a consequence of this energy dependence, for a particular ionized level, the probability of Auger emission depends upon atomic number. For example, for an ionised K level in light elements, the decay is almost 100% through the Auger channel, whereas the probability of Auger emission falls off with increasing atomic number after atomic number 12 - 15 to a value of only 30% for Zr at mass 40. Similar behaviour is seen for emission from ionized L, M, etc. levels.

The backscattering term r_M in equation [5.26] above represents the additional ionization due to electrons in the primary beam which have already undergone some interaction with the sample material but still retain sufficient energy to cause ionization of the level of interest. The backscattering term is a function of the primary beam energy, the material in which the backscattering takes place, the energy of the level to be ionized, and the angle of incidence of the primary beam. The relative effects of these contributary factors have been explored using Monte-Carlo calculations by Ichimura and Shimizu (1981). The value of r increases with the atomic number of the matrix M, and, for a given primary beam energy, decreases with the kinetic energy of the Auger electron generated. At low atomic numbers the backscattering factor increases with increasing angle of incidence, and for higher atomic number a decrease with angle of incidence is calculated. The results of the Monte-Carlo calculations Ichimura and Shimizu (1981) are summarized by Shimizu (1983) for the three angles of incidence 0°, 30° and 45° by:

$$r = (2.34 - 2.10Z^{0.14})U^{-0.35} + (2.58Z^{0.14} - 2.98) \quad (\theta_i = 0^o) \qquad [5.27]$$

$$r = (0.462 - 0.777Z^{0.20})U^{-0.32} + (1.15Z^{0.20} - 1.05) \quad (\theta_i = 30^o) \qquad [5.28]$$

$$r = (1.21 + 1.39Z^{0.13})U^{-0.33} + (1.94Z^{0.13} - 1.88) \quad (\theta_i = 45^o) \qquad [5.29]$$

Where Z is the average atomic number of the sample and U is the overvoltage, i.e. the ratio of the primary beam energy to the energy of the Auger electron. The method has been extended to give simulations of both the backscattered and the secondary electron intensities (Ding and Shimizu, 1988).

It is this strong material dependence of the backscattering factor that defeats attempts to use theoretical sensitivity factors for AES in the same way as described in section 5.4.2 for XPS. The sensitivity factor for a particular element and primary beam energy must be corrected for the sample matrix in which it is embedded, leading to the use of matrix factors in AES

5.5.3 Matrix Factors

The sample matrix dependence of the Auger signal for a given element and primary beam energy may be accommodated within the basic equation for quantitative electron spectroscopy (equation [5.2]) by the use of a matrix factor, F such that (Hall and Morabito, 1979; Seah, 1990)

$$X^A = \frac{F^{i,A} (I^A/S^A)}{\sum_j F^{i,j}(I^j/S^j)} \qquad [5.30]$$

where the matrix factor $F^{i,A}$ for Auger electrons originating from atoms of element A and travelling through the matrix i, containing j elements, describes the difference in intensity expected in emission from atoms of element A embedded in pure A and atoms of element A in the mixture i. Using equation [5.26] above together with the dependence of the attenuation lengths upon atom size and electron kinetic energy expected from the empirical relationship of Seah and Dench (1979), it is straightforward to show that, for the binary alloy AB:

$$F^{BA}(x^A \to 0) = \left[\frac{1 + r_A(E_A)}{1 + r_B(E_A)} \right] \left[\frac{a_B}{a_A} \right]^{3/2}$$

or   [5.31]

$$F^{BA}(x^A \to 1) = \left[\frac{1 + r_A(E_B)}{1 + r_B(E_B)} \right] \left[\frac{a_B}{a_A} \right]^{3/2}$$

where a_A and a_B are the atom sizes for elements A and B respectively, derived from a knowledge of density and Avogadro's Number.

For multi-component samples, it is necessary to calculate matrix factors for the various atom types in an unknown matrix. There are certain difficulties involved in this, but, as a first approximation, it is quite simple to estimate the concentrations using equation [5.2] and then calculate approximate matrix factors with which to correct the data using equation [5.31]. This procedure gives much more reliable surface composition results than using equation [5.2] in isolation but is perhaps surprisingly rarely carried out during routine surface analysis. Even with the use of a matrix correction, a degree of caution is required in the interpretation of the results of quantification. Tanuma et al (1990) point out that, for gold-copper alloys, the backscattering effect may be corrected to within a few percent, but the electron escape depth may still contribute an uncertainty of around 30%. During AES depth profiling using argon-ion erosion of the sample surface (see Chapter 6), the situation may occur in which the Auger signal intensity from a thin layer of one element is strongly affected by backscattering from a different underlying element in the substrate. In such cases, the correction proposed by Barkshire et al (1991) may be used.

The question of which are the correct sensitivity factors to use in AES is difficult are there are several published sets available in the literature (Davis et al, 1976; McGuire, 1979; Shiokawa et al, 1979; Sekine et al, 1982) with no immediately apparent method of deciding between them. The importance of using the correct set can be judged from a simple inspection of the Auger spectra given in each of these handbooks for copper. In the differential mode the ratios of the 60 eV MVV peak height to that of the LVV peak at 915 eV vary from 0.12 to 1.04. This variation is a result of the wide range of instrument characteristics and operating conditions that are used for AES. It is now known how most of these variations arise, and how data acquired on one AES instrument may be related to that acquired on another (Seah and Smith 1991). Unfortunately, there is still no satisfactory answer to the problem of which, if any, of the published datasets to use and in what way the data presented therein should be manipulated before use. The most popular of the datasets is that of Davis et al (1976). However, for accurate work the analyst should derive sets of experimental relative sensitivity factor values for pure elemental samples of the materials most likely to be encountered in a particular series of analyses, under the precise experimental conditions to be used for the analysis of the unknown samples.

As an example of the practical application of quantification of an Auger spectrum consider that shown in Figure 5.8. This is an Auger spectrum taken in the derivative mode of the exposed grain boundary surface of a commercial nickel – based alloy fractured in the spectrometer after hydrogen charging (Caceras et al 1988). The spectrum shows prominent nickel and chromium peaks with minor contributions from iron, oxygen, boron and phosphorus. The

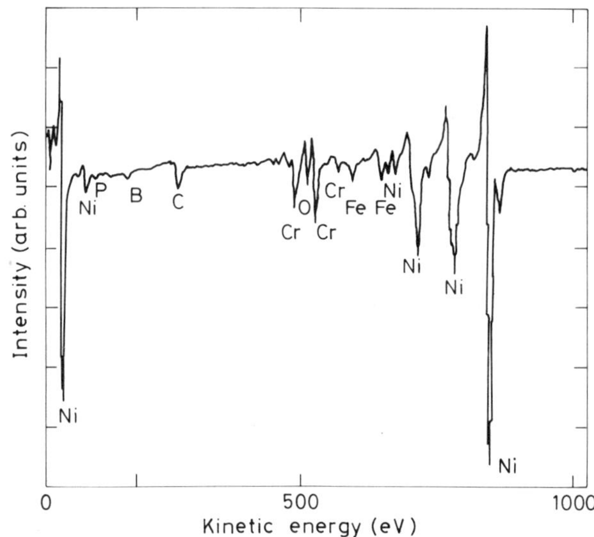

Figure 5.8 An Auger electron spectrum of the fracture surface of a commercial
nickel-containing alloy, shown in the differential mode (Caceras
et al, 1988).
Reproduced with permission, John Wiley and Co. Ltd, Chichester

measured peak to peak intensities for the peaks that will be used to quantify
the spectrum are given in Table 5.2, together with the known bulk composition
of the alloy and the various numerical factors required for the quantification. Caceras et al (1988) do not give the primary beam energy employed, so a
typical value of 5 kV will be assumed. The data were taken on an instrument
by the same manufacturer as that used in one of the reference handbooks
(Davis et al, 1976) so it is assumed that the transmission of the two
instruments is the same, and that either the spectrometer is correctly
modulated, or that the correct computer differentiation has been applied to
the data. The relative sensitivity factors (RSF) for the 5 kV beam given in
Table 5.2 are from Davis et al (1976) and, using these factors and the
measured peak intensities it is a simple matter to apply equation [5.2] to
obtain the atomic compositions in the surface region, uncorrected for matrix
effects, as listed in Table 5.2.

The results, in terms of atomic percentage composition, of this simple
analysis show nickel, iron and chromium in approximately the expected ratio
for the bulk material together with significant amounts of carbon, oxygen and
boron and a small phosphorus contribution. Carbon, boron and phosphorus are
present on the grain boundary surface in much greater concentrations than
expected from the bulk composition, probably as a result of segregation. The
presence of oxygen may be due either to contamination or to segregation from
a residual impurity level in the bulk.

It is straightforward to take the analysis of this spectrum one stage
further and include the matrix factors in the calculation, using Equation
[5.31]. In this case, the bulk composition is known and the matrix factor
for each of the elements of interest can be calculated for an average matrix
of the bulk composition. This will be quite reliable as the matrix factors
are determined by the backscattering which takes place, and this is essent-

Table 5.2 Bulk composition and results of AES analysis of the grain boundary surface of a commercial nickel-base alloy after hydrogen charging and fracture in the electron spectrometer. The column marked wt.% (a) gives the composition determined by AES if the matrix factors are ignored, and the column marked wt.% (b) shows the revised compositions after inclusion of the matrix term.

Element	Bulk wt.% (*)	Peak energy (eV)	RSF	Peak height arb. units	At.% (a)	Matrix Factor	At.% (b)
Ni	74.70	848	0.26	1.000	67.8	1.021	72.3
Fe	9.72	651	0.19	0.069	6.4	0.973	6.5
Cr	14.85	529	0.31	0.197	11.2	0.926	10.8
C	0.035	270	0.14	0.062	7.8	0.738	6.0
B	0.003	179	0.11	0.014	2.2	0.763	1.8
P	0.007	120	0.45	0.007	0.3	0.539	0.15
O	-	513	0.41	0.100	4.3	0.559	2.5

* The remainder to 100% consists of elements Al, Ti, Mn, Co, Cu and Si, which were not detected in the Auger electron spectrum.

ially a bulk effect. In equation [5.31], the atom sizes are obtained from Avogadro's number, the bulk density and atomic weight, and the backscattering factors from Ichimura and Shimizu (1981). The average atom size of this nickel - based alloy matrix is 0.225 nm with an atomic number of 27.3. Using these values the matrix factors given in Table 5.2 are obtained, which may be used to correct the original measurements to give the corrected atomic percentage compositions in the final column. Intuitively it is expected that light elements will give an enhanced Auger yield in the spectrum for the alloy, due to the extra contribution to ionisation by backscattering from the heavier elements in the matrix. Inspection of Table 5.2 shows that this is indeed the case and the corrected data show a reduction by around a factor of two for the oxygen and phosphorus concentrations, with significant reductions for boron and carbon also. Here, inclusion of the matrix effect does not greatly alter the apparent composition of the major constituents but could significantly affect the interpretation of the contributions from the important active impurity species.

STRUCTURE DETERMINATION OF INHOMOGENOUS SAMPLES

6.1 Argon Ion Sputtering

AES and XPS give surface analyses with a probe depth of a few atomic
layers. As a direct consequence of this degree of surface sensitivity, many
sample taken from the ordinary laboratory or industrial environment and
analyzed in an electron spectrometer will be found to have a surface composition consisting of primarily carbon and oxygen. This is due to the
inevitable presence of adsorbed hydrocarbons, water vapour, oxide layers,
fingerprints and so on to be found on any surface that has been exposed to
air, or handled without any special precautions being taken to maintain its
cleanliness. Analysis of these unwanted contamination layers may on occasions be interesting in its own right but generally is not very useful. To
overcome these problems requires a controlled method of removal of the
adsorbed material in such a way that these surface contamination layers are
carefully stripped away, exposing the material of interest below. Further,
samples are frequently not homogeneous in composition with depth - indeed, if
they were there would be no need for surface, as opposed to bulk, analysis -
and a means of determining this compositional variation with depth is
required for the analysis of practical samples. These requirements are met
by the technique of ion beam depth profiling.

Depth profiling by ion beam erosion of the sample is a complex topic
about which several books have been written. It is not the intention to give
a thorough review of the subject here. Instead, a general introduction is
presented with sufficient detail to enable the XPS or AES practitioner to
appreciate the requirements for good depth profiling, and to follow up
specific points in the literature where necessary.

In depth profiling, a beam of ions, typically argon in the energy range
0.5 keV to 5 keV, is directed on to the region of interest on the sample.
The impact of the argon ions causes material to be ejected from the surface
layers of the sample and, as bombardment proceeds, subsequent layers become
exposed and in turn are removed. Thus, by performing surface analysis at the
bottom of the crater so formed, the variation of the sample composition with
depth may be monitored. The sample may be continually exposed to the ion
beam, or alternatively the ion beam may be interrupted while the surface
compositions are measured. The former is appropriate where the measurement
time is small compared to the time required to erode the sample over the
depth increment of interest, as may be the case when monitoring a few intense

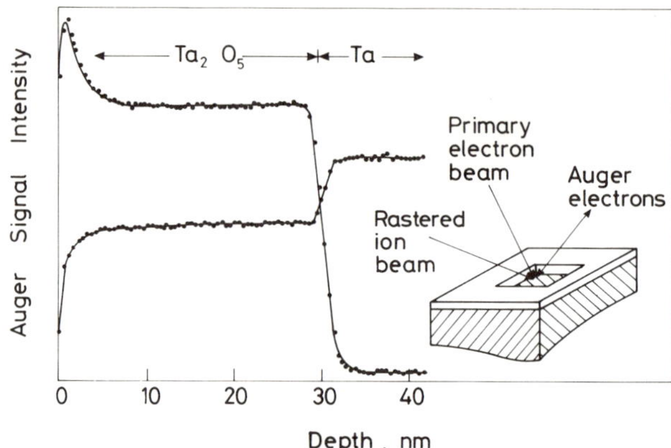

Figure 6.1 The general arrangement for sputter depth profiling using an ion
            beam in conjunction with Auger electron spectroscopy, illustrated
            by data obtained for the NPL anodically deposited tantalum
            pentoxide on tantalum depth profile reference material (Seah et
            al, 1984).

peaks by AES. For XPS a selected area analysis is usually necessary, to
avoid the need to erode a very large area of sample. Consequently, data
acquisition rates may be relatively low, and it is then appropriate to
interrupt the ion beam during the measurement.

 The depth resolution is a convolution of the sampling depth of the
surface analysis technique, which for AES and XPS is governed by the escape
depth of the electrons concerned, and the disruption caused to the true
depth-compositon profile by the mixing action of the ion beam. A factor to
be considered is that, for unambiguous results, the surface analysis must
take place only from the bottom of the crater formed by removal of the
sputtered ions. In practice this is much easier to achieve with AES than
XPS. Because of this, Auger electron spectrocopy sputter depth profiling has
historically been by far the most common method of obtaining compositional
depth profiles in working surface analysis laboratories. However, with the
development of small spot instruments, XPS is now catching up in this area.

The principles of the sputter depth profiling technique are outlined in
Figure 6.1, which shows the general arrangement of Auger depth profiling
together with the results obtained from the tantalum pentoxide sputter
profiling reference material (Seah et al, 1984). Here, the intensities of
the oxygen Auger signal at 514 eV and the tantalum Auger signal at 177 eV
are monitored during sputter-erosion by an incident argon ion beam of 2 keV
energy raster-scanned over an area of 3 mm x 3 mm. The abrupt reduction in
the oxygen signal and increase in the tantalum signal indicates the time at
which the tantalum pentoxide - tantalum metal interface is reached and, in
this case for an oxide film of 28.4 nm thickness, a depth resolution of 1.32
nm is achieved. The depth resolution is defined as the depth over which the
overlayer signal intensity falls from 84% to 16% of its original value. In
this particular material the depth of the oxide layer is carefully controlled
during production and is accurately known. This facilitates its use as a
reference standard for calibration of the rate of erosion and of the depth
resolution, both important parameters for reliable depth profiling.

Conversion of the time of exposure to the incident ion beam to a depth
scale, although of fundamental importance in depth-profiling, is not entirely

straightforward. In principle, the depth scale can be obtained from a knowledge of the flux density in the ion beam, in ions per unit area, and of the
number of atoms removed from the surface per incident ion. To a first
approximation, the depth of erosion, z, is given by the simple linear
relationship below:

$$z = (MSJ_p t)/(\rho N_A e)$$

[6.1]

where M is the mass number of the sputtered species, S is the sputtering
yield in atoms per incident ion, J_p is the primary ion current density, t is
the time of exposure of the sample to the ion beam, ρ is the density, N_A is
Avogadro's number and e is the electronic charge. The sputtering yield, S,
can be a complex function of the sample composition, and of the energy, angle
of incidence and mass of the incident ion. It is generally not known with a
high degree of certainty, and is usually the limiting factor on the accuracy
of the determination of the depth scale in a depth profiling experiment.

Sputtering is a complex process which can result in significant
modifications to the structure and composition of the surface layer under
investigation. Figure 6.2 shows schematically how atomic rearrangements can
take place due to the action of the incident ion. If the incident ion energy
is not high enough to transfer the equivalent of the surface binding energy
to the target atom then that atom will remain bound and sputtering will not
take place. Above this threshold energy Sigmund (1983) has identified three
broad regimes in which sputtering can occur.

Above the minimum ion beam energy required for sputtering, sufficient
energy is transferred to the target atoms to enable them to move in the
material. They may then either escape directly, if they have a component of
momentum normal to the surface, or they may suffer further collisions. At
low energies, known as the single knock-on regime, the primary target atom is
not able to tansfer an amount of energy greater than the binding energy in
these further collisions and so no more mobile atoms are generated. At
intermediate energies further mobile atoms are produced as a result of the
higher energy transfer during collisions. Depending upon the energies
involved, each of these atoms has a certain probability of either escaping
from the surface or undergoing collisions which may generate yet more energetic atoms in the material. This linear cascade regime represents the
processes commonly operating in most depth profiling experiments with ion

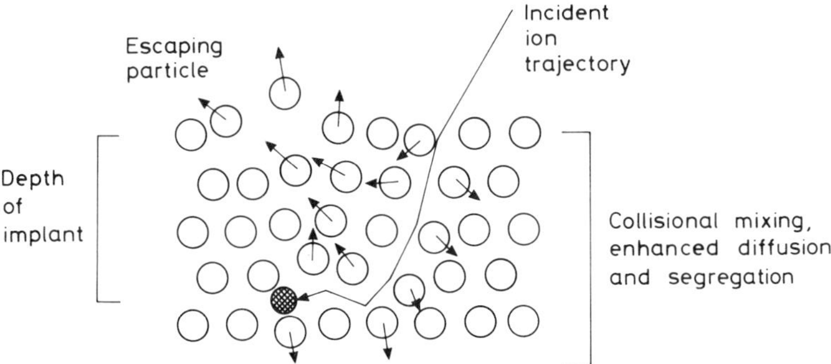

Figure 6.2 Schematic illustration of the processes that may occur during ion
beam sputtering.

beams of a few keV energy. At higher energies the spike regime operates, in
which the impact of a single ion causes a high degree of rapid particle
motion within a localised volume of the sample.

The theory of Sigmund (1977), reviewed by Sigmund (1987), was the first
approach to the problem of the sputter yield to meet with any success, and is
a good starting point for a discussion of the subject. The theory gives the
sputter yield as a function of the surface binding energy and the energy
deposited by the incident ion such that

$$S(E) = 3\ F(x,E)\ /\ 4\pi^2 C_o N U_o^2 \qquad [6.2]$$

where $F(x,E)$ is the deposited energy as a function of incident ion energy E
and distance into the target x, N is the number density of atoms within the
target, U_o is the surface binding energy and C_o is a constant approximately
equal to 1.808 A^2. Distances are measured in Ångstroms. For normal
incidence, $F(0,E)$ in the elastic collision regime of primary interest may be
expressed as

$$F(0,E) = \propto N S_n(E) \qquad [6.3]$$

where $S_n(E)$ is the energy dependence of the nuclear stopping power of the
target ion core and $\propto$ depends upon the mass ratio of the incident and target
ions. At low incident ion energies the nuclear stopping power is approximat-
ed by

$$S_n(E) = C_o T_m \qquad [6.4]$$

where T_m is the maximum recoil energy. Substitution leads directly to a
rather simple expression for the sputter yield for ion energies up to around
1 keV:

$$S(E) = 3\pi T_m / 4\pi^2 U_o \qquad [6.5]$$

In the elastic "billiard ball" approximation used here, T_m is simply given by

$$T_m = 4 m_1 m_2 E / (m_1 + m_2)^2 \qquad [6.6]$$

where m_1 is the incident ion mass and m_2 is the mass of the target ion.

This theory, applied to low energy ions normally incident upon polycrystalline elemental targets, was able to predict the sputter yields and the
variation with ion species and energy in quite good agreement with the available experimental data at the time. At intermediate and high energies the
electronic as well as nuclear stopping must be included, resulting in a
considerably greater degree of complexity. The theory is now known to
contain a number of approximations which are not quite correct, although it
remains useful in providing an understanding of some of the basic mechanisms
underlying the sputtering process.

In practice it is probably more reliable to use compilations of experimental sputter yield data than to apply uncorrected theoretical results. For
example, Seah (1981) gives predictions for all elements for sputtering by 500
eV and 1 keV argon ions, together with an extrapolation to cover the range
100 eV to 2 keV. The variation of the predicted sputter yield, with a range
of well over an order of magnitude, is shown in Figure 6.3. A more recent
compilation, giving evaluated data for selected elements over a range of ion
beam energies, is given by Chambers and Fine (1992). It is important to note
that sputter yields for compounds cannot be extracted reliably from such
compilations simply by taking weighted averages of elemental sputter yields.

Reliable reference data for sputering of compounds is sparse in the literature, and, where possible, some kind of external reference should be used.

The sputter yields show variations, often of an uncharacterised or ill-
understood nature, due not only to target atom type and projectile ion
species and energy but also to the angle of incidence, the surface topography
and crystallinity, the presence of residual gases in the UHV chamber
(especially oxygen), and impurities or minority components in the material
under study. For these reasons, tabulated sputter yields may be unreliable,
and alternative means of checking and calibrating the depth scale should be
used where possible. This may be done either outside the vacuum system,
after sputtering has taken place, or in-situ. Weight loss measurements, and
the use of interferometry or stylus-based surface profiling techniques to
measure the crater dimensions fall into the former category (Morabito and
Lewis, 1973; Laty et al, 1979) whereas quartz crystal oscillator microbalances (Lu and Czanderna, 1984), gravimetric vacuum microbalances (Akaishi
et al, 1977) and laser interferometry (Kempf and Wagner, 1984) have all been
used inside the UHV environment. Various degrees of complexity are involved
and a judgement must be made as to whether the extra information obtained
warrants the additional experimental overhead that must be carried. Probably
the most practical route to take is the use of an optical interference
microscope or a surface profiling instrument to give the depth of the crater
formed after sputtering has ceased and the sample has been removed from the
vacuum system.

As a consequence of these factors contributing to the uncertainty in the
sputter yield, equation [6.1] above is expected to give only an approximate
estimate of the depth of erosion. Further, the assumption of constant
sputtering rate is often false since as the sample composition changes so
does the density, the average mass number, and the sputtering yield.
Nevertheless, it is always useful to apply equation [6.1] in order to get a
feel for the order of magnitude involved in a particular sputtering
situation.

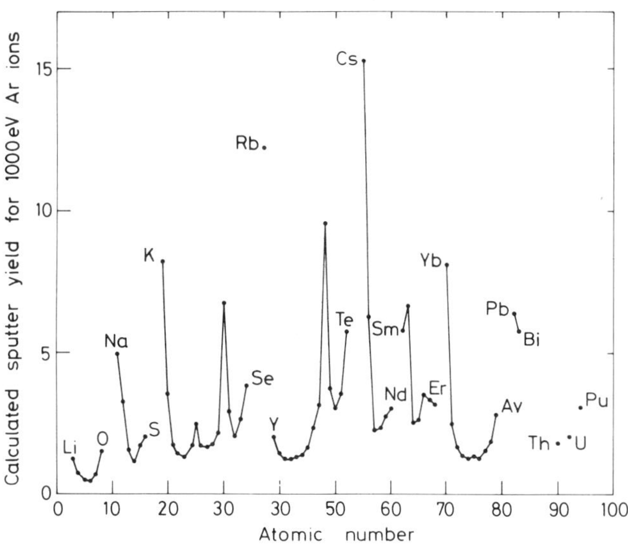

Figure 6.3 A compilation of predicted sputter yields for elements, for argon
ions incident at an energy of 1 keV, after Seah (1981).

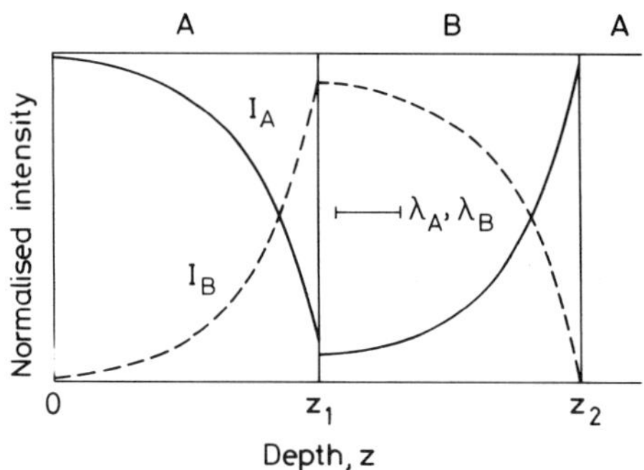

Figure 6.4 The modification to the ideal depth profile expected for a buried
layer of element B in element A due to the influence of the
electron escape depths (here assumed equal in elements A and B).

As an example of the application of equation [6.1], consider the AES
sputter depth profile of the anodically-prepared tantalum pentoxide on
tantalum layer shown in Figure 6.2. To obtain this profile, an incident beam
of energy 2 keV was rastered over an area 3 x 3 mm. Typically, an ion beam
current of 1 µA may be used for such an experiment, giving a current density
J_p of 0.11 Am^{-2} (or 11 µAcm^{-2}, a more convenient, though non-S.I., unit).
The sputtering yield for Ta_2O_5 may be estimated using the analysis of
published data given by Seah (1981), in which the calculated yields at 1 keV
incident ion energy are 1.62 and 1.77 for Ta and O respectively. Using
Seah's extrapolation to 2 keV (Seah, 1981) and taking an average yield
according to the atomic concentration gives an expected value for the yield
of 2.30. This is not likely to be very reliable, but, in the absence of any
further knowledge, it makes a useful starting point. With a density of 8200
Kgm^{-3} and the usual values for N_A and e, the time taken to sputter to the
known interface depth of 28.4 nm can be calculated to be around three
quarters of an hour, quite a feasible time for this type of experiment. To
reduce the time of the experiment the ion beam energy could be increased to
increase the sputter yield, although higher energies tend to cause a greater
degree of sample damage, or the beam current density could be increased by
reducing the raster size. This latter option requires that the electron beam
and ion beam are accurately aligned on the same part of the sample, otherwise
the measured depth profile will become distorted by crater edge effects.

In many practical cases the analysis of the composition-depth profile
would not be taken much beyond the stage of quantifying the signal intensities and defining a depth scale as outlined above. If an independent measure
of the depth of the crater is available, using one of the methods described
above, then equation [6.1] may be used to determine the sputtering yields
provided the current density and other terms are accurately known. Such
experiments using optical interference microscopy have established the
sputter yield for tantalum pentoxide at 2 keV to be 2.59, rather higher than
the value of 2.30 estimated above, although the difference between the
estimated and measured values is well within the expected limits of accuracy
of this method of estimation.

The depth resolution is usually defined as the change in depth of the
profile over which a signal goes from 84% to 16% of its value on passing
through a buried interface. The definition is based upon the true standard
deviation value for an error function profile, even though the actual profile
may not show an error function shape. There are various contributions to the
depth resolution, namely from the sample, from the analytical technique and

from the effect of the incident ion beam. The true interface width is
determined by the intrinsic properties of the sample but the other contributions can, to a certain extent, be deconvoluted from the experimental data
in order to more nearly obtain the true value (Hofmann, 1993).

In depth profiling by AES or XPS the finite escape depth given by $\cos\theta$
can make a major contribution to the observed experimental depth resolution.
The instantaneous intensity measured in such an experiment (ignoring instrumental factors) is given by

$$I = \frac{I^0}{\lambda\cos\theta} \int_0^\infty r(z)c(z)\ \exp\ (-z/\lambda\cos\theta)dz \qquad\qquad [6.7]$$

where I^0 is the measured intensity from a bulk elemental standard, λ is the
appropriate attenuation length, θ is the angle of emission relative to the
surface normal, $r(z)$ is the backscattering factor (not applicable for XPS),
$c(z)$ is the concentration profile of the element under investigation, and z
is the depth into the material. The significance of this can be readily
appreciated by considering the ideal profile of a buried layer of element B
in element A shown in Figure 6.4. Assuming an equal escape depth for both
materials, applying equation [6.7] leads to the distorted profile superimposed on Figure 6.4 (Hofmann, 1980). It is interesting to note that the
effect of the escape depth can be generalized to an arbitrary profile to give
(Iwasaki and Nakamura, 1976)

$$c(z) = \frac{I}{I^0} - \lambda\cos\theta\ \frac{d}{dz}\ (I/I_0) \qquad\qquad [6.8]$$

and therefore, provided the data are of sufficient quality, the measured
profile can be corrected for the effect of the finite escape depth of the
electrons in AES or XPS.

The various other contributions to the depth resolution, if known, may,
as a first approximation, be added in quadrature to give a resolution
function $g(z-z')$, whose effect on the measured depth profile may be
calculated via the convolution integral

$$I(z) = \int_{-\infty}^{+\infty} c(z')g(z-z')dz' \qquad\qquad [6.9]$$

In practical cases $c(z')$ is the unknown, $I(z)$ is measured and some estimate
can be made of $g(z-z')$, in which case it is necessary to apply some deconvolution procedure to determine the variation of concentration with depth.
This can either be done by numerical methods, or using an iterative procedure
in which the measured profile is compared with the calculated profile from an
assumed $c(z')$ which is subsequently improved through several cycles of
calculation and comparison until agreement is reached.

The effect of the resolution function on the quality of the depth
profiles is illustrated in the work of Olefjord et al (1990). In their work,
an interlaboratory comparison on surface analysis of thin oxide films on
aluminium was conducted under the auspices of the European Federation of
Corrosion and the European Community Bureau of Reference. Participants were
asked to measure relative XPS intensities and to perform an AES sputter depth
profile of the thin oxide film. The XPS measurements gave a value of d/λ of
1.30 ± 0.06. The film thickness, d, was known from other measurements to be
2.3 nm, so λ was determined to be 1.77 nm for Al 2p photoelectrons excited
by $MgK\alpha$ radiation and travelling through the Al_2O_3 layer. Using the known
sputter rate of Ta_2O_5 and assuming the interface was reached when the oxygen
signal decayed to 50% of its original intensity, the AES depth profiles gave
a layer thickness of 3.0 ± 0.8 nm. In this particular case, the film
thickness was approximately equal to the escape depth of the electrons used

and the discrepancy between the known and measured depths was at least
partially accounted for by the distortion effects discussed above. It was
proposed that the experimental profile be fitted by an exponential, in order
to obtain a better estimate of the depth when sputtering thin layers, as has
been done elsewhere (Kirschner and Etzkorn, 1979).

Despite the various effects intinsic to the depth profiling process
which degrade the depth resolution, it is possible, by careful control of the
experimental conditions, to minimize them in order that very high depth
resolutions indeed may be obtained. Hunt and Seah (1990) have shown depth
resolutions as low as 0.82 nm at a depth of 96.6 nm in Ta_2O_5 films sputtered
with 1.5 keV ions. This represents a depth resolution of only a few atoms,
near the achievable limit of the technique. The status of high-resolution
depth profiling is reviewed by Hofmann (1993).

Although the sputtering processes that occur in polycrystalline
elemental targets are moderately well understood, the same cannot be said for
the multi-component compound or layered structures likely to be met with in
practice. In the cases of a binary alloy, for example, it is likely that the
two components will have different sputtering yields. If the sputtering
yields in the alloy are assumed to be equal to those for the elemental state,
then it may be expected that the surface will become depleted in the component with higher yield and hence the Auger (or XPS) spectrum will give an
incorrect analysis of the true depth-distribution of the concentration. This
phenomenon is known as preferential sputtering. The situation is further
complicated by the operation of surface segregation, whereby the surface
layer of an alloy sample generally shows a different composition from that of
the bulk. This is due to segregation of the lower surface-energy constituents to the surface and is driven by the reduction in the total thermodynam-

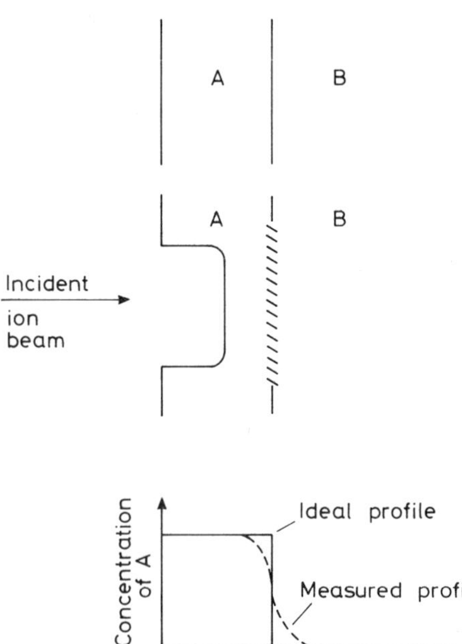

Figure 6.5 Distortion of a true atomically smooth buried interface arising
        from compositional mixing under the influence of the incident ion
        beam.

ic free energy of the system. Segregation may be enhanced by the energy
input from the ion bombardment, resulting in an increased surface concentration of one or other component of the alloy. Atoms of this type in the
surface layer then have a greater probability of being removed by sputtering,
simply as a result of there being proportionally more of them present. The
distinction between compositional modifications due to true preferential
sputtering, arising from a difference in sputter yields, and due to the
apparent preferential sputtering caused by this ion-bombardment induced
segregation is a matter of some discussion amongst practitioners of the art
(Zalm, 1988; Lam, 1988). Some authors go so far as to state that the process
conventionally referred to as preferential sputtering is often identical to
bombardment-induced segregation.

Ion bombardment induced compositional changes due to recoil implantation
and collisional mixing can affect the interpretation of sputter depth profile
data taken from samples containing buried thin layers. This is especially
significant for the study of low dimensional structures and quantum well
devices in the microelectronics industry. Although with care these unwanted
effects can be minimised, there is generally a shift and a spreading of the
depth distribution of composition in the layer. The apparent depth resolution of sputter depth profiling also depends in part on these beam induced
mixing effects. Consider the case of bombardment of an atomically smooth
buried interface between layers of material A and B as shown schematically in
Figure 6.5. As sputtering proceeds, initially atoms of the top layer, type
A, are removed. However, deeper into the material, atoms of type A are
buried further by the process of recoil implantation and eventually some
cross the boundary and penetrate into material B. At the same time, incident
ions loose energy to the lattice and cause atoms of types A and B to be
displaced from their original positions due to collisional mixing. Again,
type A atoms may be moved across the boundary into region B but also B atoms
may be relocated into material A. Thus, as sputtering continues, not only is
material removed, but the original depth distributions become smeared out and
an apparent loss in depth resolution is observed.

In addition to these mixing processes, Benninghoven (1970) has pointed
out that there is a further contribution to the loss in depth resolution
which arises simply from the statistical nature of the sputtering process.
Complete layers are not peeled away from the sample one at a time; rather, as
soon as part of the first layer has been removed, erosion of the second may
begin, and so on for subsequent layers - assuming sputtering only takes place
from the outermost exposed layer and there is no change of sputter yield.
This leads directly to an expression for the coverage, $\theta_i(t)$, of component i
of the multicomponent target after sputtering time t of the Poissonian form

$$\theta_i(t) = \sum_{n=o}^{N} \theta_{i,n} \frac{1}{n!} (t/T)^n \exp(-t/T) \qquad [6.10]$$

where T is the characteristic time taken to remove one average monolayer of
material (in terms of atoms per unit area) and parts of all layers from n = 0
to N, each with coverage $\theta_{i,n}$, are exposed. Beyond a few atom layers into
the sample this rapidly approaches a Gaussian distribution and hence a step
change in composition at depth x is predicted to have a shape corresponding
to a Gaussian integral function with a depth resolution equal to $2a^2/x$,
where a is the layer spacing (Hofmann, 1976). However, this model leads to
unphysical profiles for large ion doses, probably due to correlation between
sputtering from neighbouring sites, resulting in an overestimation of the
loss of depth resolution at higher depths (Witmaack and Schultz, 1978).

It is now generally believed that the most important contribution to the
depth resolution (apart from the intrinsic depth resolution of the probing
technique itself) comes from the collisional mixing process outlined above.

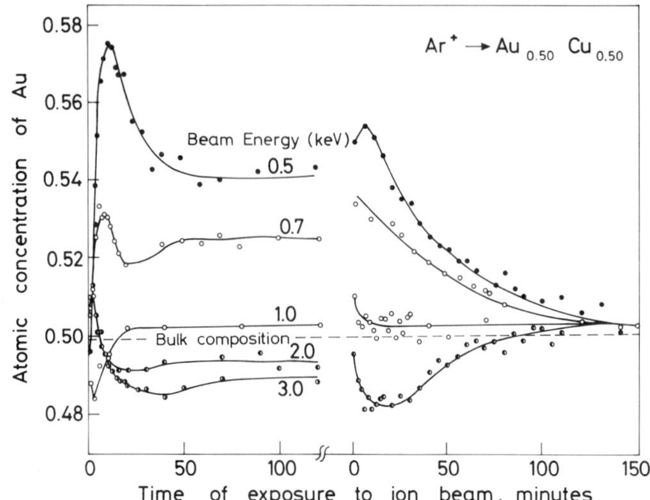

Figure 6.6 The results of Li et al (1985) for argon ion sputter depth
profiling of copper-gold alloys. The left hand side of the figure
shows the depth profiled obtained from the fresh surfaces (after
identical surface preparation) using the ion beam energies shown.
The right hand side of the figure shows the profiles subsequently
obtained using 1 keV ions incident on the corresponding surfaces
on the left - illustrating the compositional redistributions that
occured during the original profile (after Li et al, 1985)

Theoretically, the problem may be considered within the framework of a
diffusion theory (Anderson, 1979) or a transport model (Littmark and Hofer,
1980). Monte-Carlo simulations have also been used successfully (Rousch et
al, 1981). The whole theory of physical sputtering by ion bombardment has
been reviewed by Sigmund (1987). Despite the theoretical difficulties
involved and the lack of understanding of effects occuring in complex samples
it is clear that, by careful control and optimisation of the experimental
conditions used, it is possible to achieve very high depth resolutions
(Hofmann, 1993).

As an example of the type of effects that can occur even in the most
careful depth profiling experiment, consider the data shown in Figure 6.6 (Li
et al, 1985). The left hand side of the figure shows the results of
profiling a copper - gold alloy with argon ions of 0.5, 0.7, 1, 2 and 3 keV
energy. The apparent depth profile so obtained is seen to be strongly
dependent on the incident ion energy. This is interpreted as due to a
combination of preferential sputtering and ion bombardment induced
segregation. Note that for 1 keV ions a profile corresponding very closely
to that of the bulk composition (established by other techniques) is found.
After completion of the depth profiles shown on the left hand side of Figure
6.6, the samples were bombarded by the nominally-nondisturbing 1 keV ions in
order to establish the true compositional variations caused by exposure to
the ion beams of other energies. The results are shown on the right hand
side of Figure 6.6. It appears that bombardment by the low energy ions
causes a build-up of gold below the surface of the alloy whereas use of a
higher beam energy results in formation of a sub-surface gold-depleted
region. "Classical" preferential sputtering is clearly inadequate to explain
such results, which the authors interpret as being much more compatible with

interpretations based on the role of bombardment induced segregation. At present there is a paucity of reliable experimental data upon which the claims of competing theories can be evaluated, although the situation is likely to improve in the future.

A few further points, primarily of interest to the experimentalist, are worthy of mention here. First, in certain circumstances, the possibility of re-deposition of sputtered material exists and this must always be considered (Wilson et al, 1984). Secondly, there exist further contributions to depth resolution from the finite temperature and surface topography or roughness (Seah and Keuhlein, 1985). The effect of the surface topography may be dramatically reduced in some cases simply by rotating the sample during sputtering (Zalar, 1986). This facility is now provided on some commercial surface analysis instruments. Finally, it may be possible to provide automated recognition of an interface before it is reached, by employing some other signal, for instance the absorbed sample current (Hill, 1988). Coupled with computer control this can greatly reduce the time spent acquiring data of no interest in regions where the sample composition is not varying.

There exist in the literature a large number of examples of the application of depth profiling to practical technological problems, primarily in the areas of materials science and semiconductor device studies. To take just one case, the work of Clement et al (1989) on the depth profiling of $TiSi_2$ contacts deposited on GaAs is a good example of the quality of data that can be produced. They used sputtering in conjunction with AES to study interdiffusion effects as a result of annealing, while recognising the importance of preferential sputtering on the interpretation of the results. An important aspect of the work was the use of Rutherford Backscattering Spectrometry to provide a quantitative determination of the thickness and stoichiometry of the films.

In summary, while the sputtering process is reasonably well understood in the case of amorphous elemental targets, the same can not be said of the complex materials likely to be met in practice. Ion bombardment induces compositional changes by a variety of processes including collisional mixing, preferential sputtering, radiation enhanced diffusion and segregation, recoil implantation and ion-induced topographic changes. Many of these are understood in a qualitative manner only. Nevertheless, sputtering in combination with AES, or, increasingly commonly, XPS, is in routine use in almost all surface analytical laboratories, and is an extremely powerful method of obtaining information on the compositional variations with depth. While the theoretical effort intensifies in order to improve our understanding of the complexities of the processes, the experimentalists will continue to refine the techniques so that the effects of the unknown parameters are minimized.

6.2 Non-Destructive Depth Profiling by Electron Spectroscopy

Depth profiling using an ion beam is inherently destructive. Furthermore, the depth resolution is often not adequate to distinguish fine detail in the compositional depth profile of the outermost few nanometres of the sample. However, under favourable circumstances, information in this depth range can be determined by making use of various properties of the emitted electron energy distributions in XPS and AES. The methods discussed below are only applicable to laterally amorphous samples, or those which consist of a large number of randomly-arranged micro-crystallites sufficiently small that an average over a large number is obtained during analysis. If these conditions are not met, then the methods may be rendered inoperable by the effects of diffraction on the outgoing electrons (Bishop, 1991).

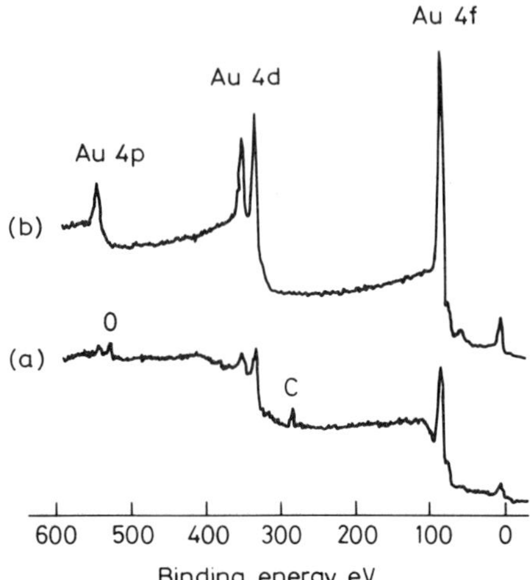

Figure 6.7 The low binding energy regions of spectra from a gold surface
(a) before and (b) after cleaning by ion bombardment. The
increase in background intensity on the high binding energy
sides of the 4d and 4f photoelectron peaks in spectrum (a)
relative to the background shape seen in spectrum (b) is typical
of the presence of a thin overlayer, in this case of
carbonaceous contamination.

6.2.1 Information from Simple Overlayers

In XPS, basic information concerning the depth location of various
elements in the sample can be obtained simply by inspection of the loss
structure accompanying the peaks in the photoelectron spectrum. An element
can be identified as being in an overlayer if its photoelectron peaks are
symmetrical with level baselines on either side and no energy loss peaks.
The absence of the loss structure implies that the electrons have not had
to undergo transport to the surface before emission, and have therefore not
had the opportunity to lose energy. On a similar argument, if the loss
structure commences some little way behind the peak, rather than immediately as would be expected for emission from a clean pure elemental sample,
then that peak is probably due to an element that is buried below an
overlayer, as electrons with small loss energies have not been able to
escape and only electrons with higher losses are emitted from the surface.

These effects are illustrated in Figure 6.7, which shows the 4d region
of the XPS spectrum of a gold sample, including the carbon 1s line, before
and after sputter-ion cleaning to remove the carbon contamination overlayer. In the as-received spectrum, the carbon peak is symmetrical, with no
associated loss structure, indicating emission is from an overlayer.
Inspection of the loss structure behind the gold 4d lines shows that,
compared to the spectrum from the cleaned sample, the presence of the
carbon overlayer causes suppression of the loss emission close to the peak,
as described above. The same effect can be seen behind the 4f line. This
approach to non-destructive depth profiling has been placed on a formal
basis by Tougaard (1986), who has shown that the variation of the
background on the low energy side of principle photoelectron peaks may be
related to the depth distribution of the emitting element with a good level
of reliability. This approach is discussed further in section 6.2.3 below.

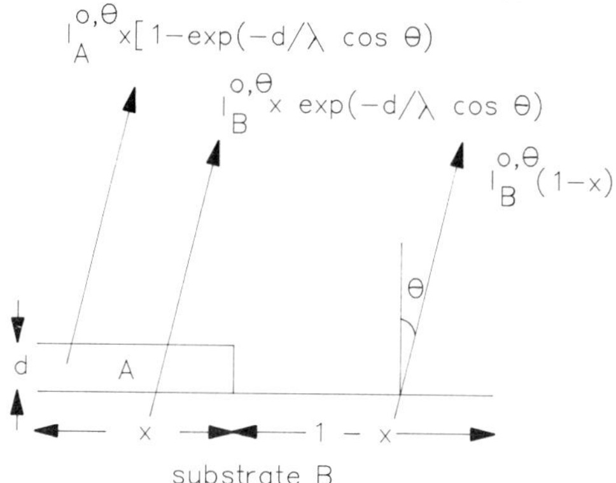

Figure 6.8 Relative intensities expected for XPS measurements of a partial overlayer of element A of coverage x, on a substrate of element B.

An alternative method of obtaining non-destructive depth distribution information from XPS spectra makes use of the energy dependence of the attenuation lengths of photoelectrons. This method is suitable for elements with lines at widely spaced energies in the spectrum. From the "universal curve" for attenuation lengths (see Chapter 2) it can be seen that, for electron kinetic energies above around 100 eV, higher energy electrons are attenuated less than lower energy electrons. Thus, for a buried element, the higher energy peak in the spectrum will appear relatively stronger than the lower energy peak, when compared to the pure clean elemental spectrum, and similarly the lower energy peak will appear relatively stronger for an element in a surface layer. As an example, the method has been used by Bernstein and Grepstad in a study of the effects of different etching treatments on the outermost surface layers of GaAs crystal slices (Bernstein and Grepstad, 1989). Data on relative peak intensities are available for XPS (Wagner et al, 1979) but, as with all reference data, care must be taken to ensure the experimental measurements are taken under the same conditions as the reference data, otherwise uncharacterized relative instrument energy-intensity response functions could make the method unreliable. If this technique is to be used often, the analyst may find it worthwhile to make his or her own measurements of relative peak intensities for the elements of interest.

Often, the analyst is faced with a sample that is known not to be a simple homogeneous mixture. Typically, the surface to be analysed may consist of a contamination layer, or an evaporated overlayer on a homogeneous substrate, or a sample in which the composition of the outermost layer is different from the bulk as a result of surface or grain boundary segregation. In such cases, a rather different approach from that outlined above must be used. Consider the situation shown schematically in Figure 6.8 in which a partial overlayer of fractional coverage x of element A covers a substrate of element B. The emitted spectrum contains three contributions: that from the overlayer, that from the exposed part of the substrate, and that from the covered part of the substrate. Thus, the total substrate emission is given by

$$I_B = I^o_B (1-x) + xI^o_B \exp(-d/\lambda_A(E_B)\cos\theta) \tag{6.11}$$

and emission from the overlayer is

$$I_A = x \, I^O_A \, [1 - \exp(-d/\lambda_A(E_A)\cos\theta)] \qquad [6.12]$$

In practice, ratios of intensities are measured and, for small coverages of monolayer thickness overlayers, a monolayer matrix factor Q_{AB} can be defined (Seah, 1990) such that

$$x = Q_{AB} \, \frac{I_A/I^O_A}{I_B/I^O_B} \qquad [6.13]$$

where for XPS:

$$Q_{AB} = \frac{\lambda_A(E_A)\cos\theta}{a_A} \qquad [6.14]$$

and in AES:

$$Q_{AB} = \left[\frac{\lambda_A(E_A)\cos\theta}{a_A}\right]\left[\frac{1 + r_A(E_A)}{1 + r_B(E_A)}\right] \qquad [6.15]$$

These equations can, in principle, be extended to incorporate a variety of more complex overlayer and substrate structures and compositions (Fulghum and Linton, 1988).

6.2.2 Structural Information from Angle-dependent XPS

If the spectrometer is equipped with a facility to restrict the solid angle from which photoelectrons are collected, then the technique of angle-resolved XPS may be employed to give non-destructive information. As we have already seen, in AES and XPS the information depth is given approximately by the Beer-Lambert law

$$I(\theta) = I^O \exp - (z/\lambda\cos\theta) \qquad [6.16]$$

where $I(\theta)$ is the intensity of emission at angle θ measured relative to the sample normal originating from depth z in the sample and λ is the appropriate electron attenuation length. The depth sensitivity can clearly be varied by changing θ. Therefore, by taking spectra near normal emission (large information depth) and near grazing emission (small information depth) and comparing relative intensities, information on the relative depth distributions of the various elements or chemical states contributing lines to the spectrum can be obtained. For example, a spectral line due to a surface species will be enhanced at grazing emission relative to a line from a buried species. Useful information concerning depths up to around 3λ can be obtained, typically up to a maximum of around 10 nm. To extract more detailed information from the spectra may in practice be quite difficult. Ideally, the analyst would like to be able to reconstruct the true composition depth profile of the sample from a series of spectra taken at different emission angles. If the sample is known to consist of a single overlayer of element A on element B then the thickness of the overlayer is simply given by (Fadley, 1984)

$$d_A(\theta) = \lambda_A \cos\theta \, \ln \left[1 + \frac{I_A}{I_B} \cdot \frac{I_B^O}{I_A^O}\right] \qquad [6.17]$$

using the nomenclature above, so long as the peaks from A and B are not greatly separated in energy. For plane surfaces and overlayers, intensity ratios I_A/I_B for various values of θ can be used to determine d_A to an accuracy primarily limited by the knowledge of the attenuation length, and the roughness of the surface (Yan et al, 1989). The method is readily extended to include multilayers, again provided the structure is known. This simple approach has been used successfully in studies of lubricant films applied to magnetic recording media (Swami, 1989), and superconducting niobium oxide films (Darlinski and Halbritter, 1987), among others. A review of ways in which angle-resolved XPS data from simple systems may be treated using variations on equation [6.17] is given by Fulghum (1993), and a comparison of angle-resolved XPS with SIMS depth profiling (see Chapter 7) for simple assumed exponential compositional profiles is discussed by Ebel et al (1991).

If, as may more usually be the case, there is only limited a priori knowledge of the in-depth structure of the sample, then the data set of peak intensities vs angle must be treated mathematically in order to yield the depth profile. Unfortunately, the inversion of angle-dependent XPS data is one of a class of mathematically ill-posed and ill-conditioned problems. Standard numerical techniques such as the Laplace Transform cannot be used directly to generate meaningful depth profiles, as small amounts of noise in the experimental data can result in unphysically large variations in the transformed result. For success, the transform must be constrained in some way, either using a damping method or employing knowledge about the overall form of the profile extracted from elsewhere. This can result in restrictions on the allowed solution that may preclude certain types of profile, for example those with more than one point of inflection or with rapidly varying profiles with depth. For the complex technological samples met with in practice, these restictions can prove serious. It appears possible to overcome these restrictions by making use regularisation techniques (Tyler et al, 1989; Baschenko and Nefedov, 1990). However, the degree of regularisation chosen can result in an overconstrained fit to the data.

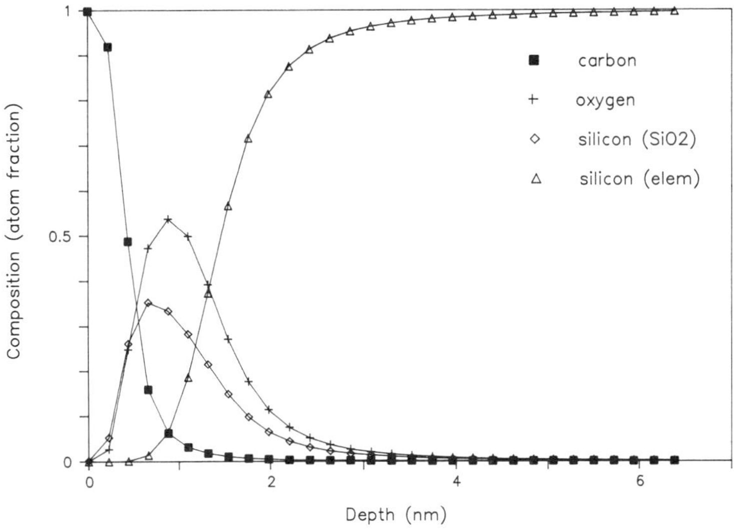

Figure 6.9 Reconstruction of the depth profile of a thin, contaminated, oxide layer on a single-crystal silicon surface. The XPS data are from Tyler et al (1989), and the depth profile is obtained using the maximum entropy method of Smith and Livesey (1992).

An alternative to these inverse tranform methods is to make use of the forward transform. Several authors have attempted to calculate trial data from model sample structures, iterating the structure until some degree of agreement is reached between the calculated and the measured data (Watts et al, 1986; Hazell et al, 1985). This approach suffers from the serious problem that, within realistic levels of experimental noise on the measured data, there are many possible structures which can produce data in apparent agreement with the measurements.

Recent work has shown that the ambiguity of solutions inherent in the forward transform approach may be removed by using maximum entropy data processing, without overconstraining the solutions (Smith and Livesey, 1992). This method automatically returns the reconstruction with the least information content – i.e. the maximum entropy – consistent with the level of experimental noise in the data, and as such provides a logically consistent approach based soundly on Bayesian statistical theory. The method has the added advantage that, being based upon a forward transform, it is easy to include the effects of elastic scattering on the data using the parametrizations of Werner (1991) or others that may be developed subsequently. Elastic scattering has been included in the methods based upon the Tichanov regularisation described by Baschenko and Nefedov (1990) (Baschenko and Nesmeev, 1991), but this requires the use of Monte-Carlo calculations to determine the appropriate parameters.

Figure 6.9 shows a depth profile reconstructed from angle-resolved XPS data using the maximum entropy data processing method (Smith and Livesey, 1992). The reconstruction is from the data of Tyler et al (1989), who give the intensities of the C 1s and O 1s lines, together with the Si 2p intensity resolved into components due to the elemental and oxide forms, at 3 different angles of emission. These 12 data points give a physically realistic reconstruction over 30 depths, showing a carbon overlayer (presumably from adventitious hydrocarbon contamination) over an approximately SiO_2 film on elemental silicon. The result is smoother (i.e. less structure, therefore higher entropy) than the reconstruction obtained using the regularization method applied in the original paper of Tyler et al (1989).

There has been much recent research activity in this field which is now bearing fruit. An interesting recent development is the application of angle-dependent XPS methods to studies of the surfaces of liquids (Baschenko et al, 1993). Although posing severe experimental difficulties, this has the advantage that the liquid surfaces are absolutely smooth and therefore may make ideal samples for the evaluation of the different data processing methods. Clearly, in due course further examples of the application of these methods will be seen in the literature, and eventually their implementation on commercial surface analysis data systems can be expected.

6.2.3 Peak to Background Ratio Methods

The basic idea behind methods of non-destructive profiling based on meas-urements of peak to background ratios is that electrons originally excited at various depths in the sample loose energy on their way to the surface. The peak shape and the background behind it in the emitted spectrum are therefore distorted by an amount dependent on the path length travelled by the electrons. Hence, they depend upon the depth composition profile of the element contributing to the peak. For XPS, this effect is automatic-ally included in the algorithms proposed by Tougaard (discussed in Chapter 4 section 4.5) for accurate background subtraction. The relationships for background correction given in section 4.5 are valid for a homogeneous distribution of material in the sample. However, Tougaard (1986, 1986) has pointed out that the theory of the background shape can be extended to provide a method for the determination of composition depth profiles. The method uses the quantity $D = A_p/B$ where A_p is the peak area and B is the increase in background signal behind the peak. D is found to be approximately constant for XPS peaks from homogeneous transition elements, independent of concentration, peak shape and peak energy. It is, however, strongly dependent on the distance R travelled by the photoelectron before escaping from the surface. The dependence of D on R is given by Tougaard (1986) for buried layers of transition metals and therefore, from one measurement on a single spectrum, the depth of the buried layer may be deduced provided the depth is not too small (Tofterup, 1989). Vignes et al (1985) show experimentally that similar measurements made on Auger electron spectra may also be used to determine information on the depth distributions of elements in the sample, although their method relies on calibration from known structures and is not general.

The method has been extended to include information from the full background shape, rather than at one single point (Tougaard, 1988) in order to give greater reliability, and to enable exponentially-varying depth distributions to be characterised. Tests on the validity of this approach made on gold-nickel alloys with unknown depth composition profiles (Tougaard et al, 1991) show that the method gives an improvement of over an order of magnitude accuracy, compared to the usual approach of assuming homogeniety and applying sensitivity factors for quantification. It is interesting to note that similar arguments may be made concerning the extraction of non-destructive depth profile information from Auger electron spectra (Tofterup, 1986; Peacock, 1985; Sickafus, 1977; Langeron et al, 1984)).

6.2.4 Studies of Rough or Random Samples

The discussion of non-destructive depth profiling given above has assumed that samples are perfectly flat in structure. In practice, of course, for solids this is not the case. No matter how carefully prepared, even the best single crystal slice of semiconductor will exhibit some degree of roughness, and many samples are very far from this ideal.

In general, the effects of surface roughness do not appear to be as serious as may be expected. The important parameter appears not to be the roughness as determined in the usual way as a centreline average, or an r.m.s. deviation from a flat plane. Rather, it is the mean angular variation in the surface, as variations in the angle of the surface presented to the analyser are equivalent to a corresponding degradation in the angular resolution of the spectrometer used for the measurement. Since most angle-dependent XPS measurements are made with analyzer angular resolutions of around 3° or worse, it is clear that some variation in mean surface angle is tolerable. Furthermore, in the area-averaging instruments typically used for XPS, gross defects such as scratches can be tolerated so long as they only occupy a small fraction of the total area under analysis. Nevertheless, it is clearly necessary to achieve as high a degree of surface finish as possible for the particular sample type. For metallic samples, polishing down to 0.25 μm with diamond paste to give a mirror finish has been found to be adequate.

For samples with a greater degree of roughness than this, the degree of angle averaging, even in a spectrometer with good angular resolution, becomes more pronounced. In the limit of an infinitely rough, or random, sample, angle averaging is complete. Such samples show no variation in apparent composition as a function of emission angle of the photoelectrons, and measurements at all angles are equivalent. This is necessarily so for powdered samples, however, in certain specific cases, particularly that of catalysts, this randomness can be exploited to give structural information.

The analysis of catalysts often involves determining the relative amounts of material present in different phases, typically a finely dispersed powder on a support material. This is an extreme case of a non-homogenous material, and as such the usual methods of quantitative XPS cannot be applied. Kuipers et al (1986) show how, with the simple assumption of randomness, XPS can be used to analyse these materials to give information such as particle size and the coverages of active compounds. By starting with the principle that, for truly random samples, all possible surface orientations are equally probable, they show that for such samples the XPS signal intensity for a supported phase which is present as equally sized but arbitraliy shaped convex particles is deter-mined by the surface to volume ratio of the particles. The surface to volume ratio indicates the degree of dispersion of the active constituent of the catalyst, and is critical in determining its performance. It appears from this work that XPS "looks" at the surface in the same way as the reactants. This partly accounts for its extensive use in this area.

The work of Kuipers et al (1986) has been extended by ten Berge et al (1989) to allow the determination not only of the average particle size in heterogenous catalyst, but also the particle size distribution. The work appeared to be validated by transmission electron microscopy but contains a large number of fitting parameters and may be difficult to apply to anything other than well characterised systems.

CHAPTER 7

TRENDS IN SURFACE ANALYSIS

7.1 Introduction

The previous six chapters of this book have provided an introduction to
the analysis of solid surfaces by the principle methods of practical electron
spectroscopy, namely X-ray photoelectron spectroscopy and Auger electron
spectroscopy. Overviews have been given of the the theoretical background to
the techniques, of the kind of experimental apparatus used, and of the means
available for extracting information and knowledge from the data produced by
the instruments. However, neither of these techniques can yet be thought of
as fully mature. A continuous process of improvement and refinement is under
way, in the sensitivity, ease of use and range of applicability of the experimental methods, and in the degree of true theoretical understanding under-
pinning the analysis of the results and the derivation of numerical quantities of interest from them. Trends can be discerned in the direction and
likely outcome of these processes. These are discussed in section 7.2 below.

Of course, AES and XPS do not exist in isolation as the only possible
methods of surface analysis. From the earliest days of surface analysis, the
benefits of a multi-technique approach have been recognised. The techniques
have been used alongside other methods in attempts to obtain full pictures of
the composition, structure and chemistry of the surface and near-surface
regions of the samples of interest. In particular, techniques employing
ions, rather than electrons, as the detected species have proved complementary to AES and XPS, although they are by no means the only species that may
be used. To place the contents of this book in the wider context of surface
analysis at large, the final section of this chapter is devoted to a brief
comparison of AES and XPS with the more common alternative techniques.

7.2 Trends in Surface Analysis by XPS and AES

Whatever the spatial resolution, sensitivity, spectral energy resolution, ease of use, or other desirable parameter of a particular instrument or
technique, it is a truism to say that it is not enough. In commercial
laboratories (and commercial offshoots of higher education institutes), AES
and XPS, particularly when operated as part of a larger analytical facility
offering a range of techniques, will be subject to stringent processes of
cost-benefit analysis. The desire for continuous improvement in sample
throughput and overall efficiency provides the drive for the innovative
designs seen in the instrumentation from the principle manufacturers. The

manufacturers' need for product differentiation in a competitive market also
contributes to this process. Many of these improvements are in aspects of
instrumentation discussed in Chapter 3, and particularly concern sensitivity,
spatial resolution and energy resolution. Enhanced data processing using
validated algorithms and, possibly, expert systems is another area in which
trends can be identified.

In any one instrument design, the sensitivity and the spatial and energy
resolutions are intrinsically linked. In AES, reducing the beam spot size to
obtain maximum spatial resolution on the sample surface will inevitably lead
to a reduction in beam current and a corresponding reduction in sensitivity,
if not from fundamental design limitations then from a desire to avoid intol-
erable radiation damage to the specimen. However, using ultra-high vacuum
electron microscope technology coupled with magnetic field parallelizers to
recover the Auger electrons, there seems to be no fundamental reason why the
spatial resolution of AES, when governed by the primary beam diameter, should
not be pushed to the limit of single atom analysis defined by Cazaux (1987).
Such a high resolution experiment may require special sample geometries to
avoid the backscattering effect normally encountered as a possible contribution to resolution limitation in conventional apparatus.

In XPS, the achievable spatial resolution has undergone dramatic improvements recently. A significant contribution here has been the development of
parallel acquisition imaging instruments, as described in Chapter 3. The
possibility of acquiring data from all points of the image simultaneously,
rather than pixel-by-pixel using a raster scan, is very attractive and must
surely act as a pointer to the future. The principle manufacturer of instruments of this type has claimed an image resolution as high as 2 μm, with
reasonable acquisition times of only a few minutes. It remains to be seen
how this is born out in practical use. Nevertheless, it represents an
improvement of two orders of magnitude over what was considered good XPS
spatial resolution only a few years ago, and, furthermore, rivals the limit
achievable with optimum designs of current-technology zone plates used in
conjunction with synchrotron radiation (Lodders et al, 1992).

As well as the imaging electron optics, a key factor in allowing
progress in small spot- or imaging-XPS has been the availability of a
sufficient flux of X-rays onto the sample area to be imaged. This has been
acheived by significant improvements in monochromator design, making optimum
use of the focussing properties of the curved monochromator crystal. A
further development has been the use of position sensitive detectors to
enable collection of all electrons across the dispersion plane of the spectrometer. These detectors improve the performance of both AES and XPS instruments. In the simpler sytems, the position sensitive detector may be three
or possibly five single channel electron multipliers placed across the output
slit, while more sophisicated systems may use a single truely position sensitive device, possibly based on a resistive anode encoder or a system of
channel plates followed by some kind of photoelectric detector.

7.3 A Comparison of AES and XPS with Alternative Methods of Surface Analysis

There are many texts available that compare and contrast the multitude
of surface analytical techniques that can be devised by varying the possible
combinations of excited and detected species (electrons, photons of various
energies, ions, neutral species, etc). An excellent, and exhaustive, review
is given by Riviere (1990). Here, the focus will be on the analytical techniques most likely to complement XPS and AES in practical laboratory situations where it is desired to elucidate the surface composition, chemistry and
structure of an unknown sample. The methods with the highest capacity to
generate useful information in such circumstances in practice are the ion

Table 7.1 A comparison of the principle characteristics of XPS, AES and complementary surface analytical techniques. y = yes, n = no, and (y) means the possibility is available but not is normally the prime reason for using the techniques. The figures for resolution, sensitivity, etc., can be very instrument and application-dependent; those given here represent typical values obtainable on good modern equipment.

			Technique			
Characteristic	XPS	AES	SIMS	ISS	EPMA	IR
Probe species	X-ray	e^-	ion	ion	e^-	IR photon
Detected species	e^-	e^-	ion	ion	X-ray	IR photon
Elemental data	y	y	y	y	y	n
Chemical data	y	(y)	y (static)	n	n	y
Quantification	y	y	y, with close standards	(y)	y	n
Sensitivity	0.1%	0.1%	ppm-ppb	few% to 0.001%	0.01%	0.1 ml
Depth sensitivity	<1 nm	<1 nm	1 ml (static)	1 ml	10 nm	1 ml
Depth profiling	y	y	y	y	n	n
Depth resolution	2-5 nm	2-5 nm	>2 nm	>2 nm	1 μm	-
Spatial resolution	few mm - 10 um	100 nm	100 nm	0.1-1 mm	1 μm	few mm

beam techniques of secondary ion mass spectrometry (SIMS) and ion scattering spectroscopy (ISS), coupled with the more conventional non-UHV probes such as electron probe microanalysis (EPMA) and reflection-absorption infra-red spectroscopy (RAIRS). These are briefly summarized in turn below. The descriptions are intended to be superficial, for the purposes of comparison only; for more details the reader is referred to the literature. A summary of the principle features of each of the techniques discussed is given in Table 7.1.

7.3.1 Secondary Ion Mass Spectrometry

In AES or XPS depth profiling using an ion beam as described in Chapter 6, the ion beam erodes the sample and the electron spectroscopic technique is used to interrogate the freshly-revealed surface. SIMS is conceptually attractive as the ion beam is used as both an excitation probe, and as a means of removing material. As its name implies, the objective is to analyze the secondary ions removed from the sample surface in order to obtain information on the original surface composition and chemistry. SIMS is

carried out using several possible experimental configurations, each designed
to optimize a different aspect of the technique. This divergence has
resulted in two separate fields of SIMS analysis, namely static-SIMS and
dynamic-SIMS. Several reference works are available, giving excellent
reviews of the various aspects of SIMS (see, for example, Briggs and Seah,
1992; Wilson et al, 1989)

In static-SIMS, the philosophy of the experiment is to be as gentle to
the sample surface as possible. The kinetic energy carried by the incident
ions is dissipated below the sample surface, resulting in the desorption of
possibly quite large molecules and molecular fragments from the surface layer
of the sample. These species are then analyzed and detected using an
analyzer optimized for high sensitivity, with mass resolution an important,
but secondary, consideration. Time-of-flight analysis is often used, with
either a focussing dispersion element, or a reflectron geometry. Analyzers
can be of such a high sensitivity that an incident ion beam flux calculated
to remove one monolayer of material in 24 hours or more can give a sufficiently high signal for the acquisition of a spectrum from a sample surface in
a few minutes. This extreme of the SIMS technique is very well suited to the
analysis of polymer surfaces, and handbooks giving typical static SIMS
spectra from a range of polymer materials are available (for example Briggs
et al, 1989).

Static-SIMS can provide valuable information to complement XPS investigations of surface chemistry. Whereas XPS is sensitive to the local
electronic environment around the emitting atom, static-SIMS gives chemical
bonding information directly from an analysis of the molecular clusters seen
in the spectra. The surface sensitivity of static-SIMS is greater than that
of XPS, the detection limits are orders of magnitude better, and the analysis
may be carried out in an imaging mode with greater spatial resolution than
usually obtained in XPS. Unfortunately, secondary ion yields vary by many
orders of magnitude between species and host matrices. As a consequence, the
technique is not quantitative, although useful comparisons of peak intensities across a range of similar samples can be made. XPS and static-SIMS have
been successfully used in combination in a wide range of technological
investigations, ranging from the study of materials bio-compatibility to
applications in adhesion and wear

Dynamic SIMS can be thought of as the intellectual opposite of static-
SIMS, with an objective of inducing a high degree of surface damage, albeit
in a controlled manner. The apparatus is optimised for a maximum etch rate,
and a high mass resolution in the spectrum of species detected, and the
experiment is performed by monitoring the intensity of particular ion species
as a function of time of exposure to the incident ion flux. Conversion of
time of exposure to a meaningful depth scale is difficult, as discussed in
Chapter 6 for AES and XPS depth profiling; it is normally accomplished by
means of comparison with standards. Magnetic-sector mass spectrometers, with
ion energy analysis, are often used for dynamic-SIMS. Such devices are not
generally compatible with other methods of surface analysis, and are operated
as dedicated single-function instruments. This contrasts with the analyzers
used for static-SIMS, where it is not unusual to find XPS, for example, built
into the same instrument.

Dynamic-SIMS has been adopted as the technique of choice in many
analytical laboratories in the microelectronics industry, where sensitivity
and spatial resolution are overriding factors. For many elements in silicon,
sensitivities in the range of parts per million to parts per billion are
found. In analytical chemistry terms, this corresponds to the detection of
picomoles of material - something hard to achieve with other techniques, and
virtually impossible if spatial resolution, laterally or in depth, is
required.

If XPS is the natural partner of static-SIMS, then AES has an equivalent relationship to dynamic-SIMS. AES can generally be performed at higher spatial resolution, and is quantitative if carried out with care. Dynamic-SIMS has the greater sensitivity and can cope with a greater range of signal intensities, but, without standards of very similar composition to the sample under investigation, it is not quantitative. Often, the compositional depth-profiles obtained from dynamic-SIMS are used as a diagnostic in device failure analysis, where standards in the form of working devices are available for comparison and the problem of quantification does not arise.

The discussion above has concentrated, in an introductory manner, on the two extreme forms of SIMS analysis. In fact, probably the most widespread form of analyzer used for SIMS is the quadrupole mass spectrometer. This type of analyzer can be compact, compatible with other forms of UHV surface analysis equipment, and adapted for either static (or, for the smaller instruments, quasi-static) or dynamic SIMS measurements. The performance is not so great as the more sophisticated dedicated instruments, but the flexibilty of integration into a multi-technique environment can more than make up for this. Generally, quadrupole mass spectrometers for SIMS analysis are operated at around one mass unit mass resolution and are restricted to a range of at most 1000 mass units, with a sensitivity of possibly several orders of magnitude worse than the best dedicated instruments. Nevertheless, they are relatively simple and reliable in operation, and have often been built into combined AES/XPS/SIMS surface analysis systems.

7.3.2 Ion Scattering Spectroscopy (ISS)

In ISS, ions are used as the incident and detected species. However, unlike SIMS, it is the ions originally in the incident beam and scattered from the surface that are detected, and they are energy- rather than mass-analyzed. The technique is sensitive primarily to the outermost atomic layer of the sample, giving an even greater surface sensitivity than static-SIMS.

Interpretation of the spectra proceeds by relating the energy losses suffered by the incident ions to the masses of the surface atoms they have interacted with. For the low energy ions in the 0.5 - 5 keV range used for ISS, binary collision theory based on Newtonian mechanics is sufficient to describe the principle spectral features. Scattered intensities are measured as a function of scattering angle as well as energy loss, as this can be exploited to give structural information, particularly from ordered layers on single-crystal surfaces. The favoured incident beam is helium, for its low mass, with the helium-3 isotope sometimes preferred over the more abundant helium-4 for the increase in sensitivity which is obtained. Neon or argon ions are also used, where higher mass resolution is required for the analysis of samples containing high mass number elements, but with the penalty of a greater amount of surface damage.

The ISS experiment can be carried out using a simple electrostatic analyzer of the type more commonly applied to AES or XPS. All that is required is a change of polarity from negative to positive deflection voltages. However, an important aspect of ISS is the extreme surface sensitivity. In order to exploit such sensitivity, it is necessary that the outermost surface layers are not destroyed during the analysis. This means that instrumental sensitivity of similar order to that required for static-SIMS should ideally be used. Time-of-flight analyzers have been used for ISS, but they are generally large and complex pieces of apparatus which cannot be moved around inside the UHV system. Sensitivity is therefore obtained at the loss of the potential for structure determinations.

The principle disadvantage of ISS is that it does not directly give quantitative compositional information. Intensities can be compared with those obtained from pure element standards, but ISS would not normally be used solely as a means of determining surface compositions. Instead, it is more likely to be exploited in cases where it is desirable to be able to follow changes in the outermost surface layer of the sample as a function of time, or of some kind of process. The principle applications are in single-crystal studies of surface reactions, for example in catalysis, but it has potential in other fields such as microelectronics. An elegant use of the technique is as a highly sensitive monitor of compositional changes during ion-beam depth profiling, with the same beam in use for sample erosion and ISS measurements.

The surface sensitivity of ISS can prove to be a barrier in its application to the analysis of more technologically-relevant samples where some kind of contamination layer is often present. Nevertheless, it is a useful technique, and can be very complementary to XPS or AES analysis.

7.3.3 Electron Probe Microanalysis (EPMA)

Electron probe microanalysis (EPMA), or energy dispersive analysis of X-rays (EDX), is a common feature of analytical scanning electron microscopy (SEM). EDX detectors are often fitted to AES instruments to provide a "bulk" analysis to contrast with the surface analysis by AES from the same point on the sample surface, when addressed by the same electron beam. However, a body of work is emerging which demonstrates that the electron microprobe can be used for the analysis of thin films, perhaps below 10 nm in favourable cases, and can therefore encroach on the territory previously reserved for UHV surface analysis.

In conventional EPMA, the X-ray fluorescence produced by the SEM electron beam is detected in a solid state silicon-lithium drift detector. Quantitative analyses of the bulk composition of the sample are made by comparing the measured X-ray intensities with those obtained from elemental standards, after correction for atomic number, absorbtion and fluorescence in the sample (the so-called ZAF correction - the equivalent of the matrix correction in AES). As an alternative technique, a wavelength-dispersive crystal spectrometer may be used. These two types of instrumentation give rise to the acronyms EDS, for energy-dispersive spectrometry, and WDS, for wavelength-dispersive spectrometry. EDS has the advantage that it is rapid, because the spectra are collected in parallel. Unfortunately, for experimental reasons, the atomic number range of elements detected is usually restricted to those above fluorine or sodium, and the energy resolution is relatively poor. WDS is slower, because the spectrum has to be scanned across the detector, but the energy (or wavelength) resolution is very much better, enabling structure in the lines to be detected and interpreted, and elements down to carbon can be quantified. A well-equipped analytical SEM will incorporate both types of detector.

Under normal operating conditions, the mean depth of generation of the X-ray fluorescence is around 1 μm. However, by reducing the electron beam energy from typically 25 keV to, for example, 5 keV, this depth may be reduced to less than 100 nm, depending on the energy of the X-rays concerned. Obviously, the incident beam energy has to be above the threshold for X-ray excitation from the elements of interest, and this can be a limitation in some cases. With a mean X-ray generation depth of less than 100 nm, a substantial amount of the final signal detected will originate in the outermost few nanometers of material, giving rise to the possibility of detecting (and quantifying) films of less than 10 nm thickness.

The physics of the X-ray generation process are sufficiently well understood for accurate models of the expected intensity ratios from known sample structures incorporating thin layers to be available (Heinrich and Newbury, 1991). A number of computer programs have been written which model the X-ray intensities expected from known structures, or provide some degree of unknown structure determination from measured intensities. Of these, the program GMRfilm (Waldo et al, 1993), available through the Microbeam Analysis Society (MAS Software Library, J.F. Mansfield, University of Michigan, Ann Arbor, MI), is particularly useful in that it allows both the calculation of X-ray intensity ratios from known structures and the determination of unknown structures (within certain limits) from the measured intensity ratios, with a choice of physical models according to the preference of the user. The degree of structural information obtained by such means can be extended by the use of X-ray intensity measurements as a function of primary beam energy. This is an area of current active research in which rapid progress can be expected.

7.3.4 Reflection-Absorption Infra-red Spectroscopy (RAIRS) and Diffuse Reflection Infra-red Fourier Transform Spectroscopy (DRIFTS)

Optical spectroscopy of surfaces is attractive in that it offers the possibility of surface analysis without the requirement for a vacuum system. Consequently, problems of sample size, vapour pressure and volatility become much less important. It is possible to contemplate surface analysis of liquids without the very specialized equipment used in the few electron spectroscopic studies reported in the literature. In-situ analysis of, for example, a surface exposed to a reactive gas is also possible.

Infra-red (IR) radiation is the most suitable, and most established, optical probe of surface composition and chemistry. IR spectroscopy works by detecting the wavelengths at which energy is absorbed as the beam passes throught the sample under investigation. Chemical bonds are excited into different vibrational modes at specific energies in the IR spectral region, and consequently absorb radiation at these wavelengths. Measurement of the spectrum of wavelength dependent absorbances therefore gives a probe for the presence and relative amounts of different chemical bonds. In the standard transmission geometry, IR spectroscopy is a well known and widely used method of bulk chemical analysis.

Surface sensitivity can be achieved in IR spectroscopy by reflecting the IR beam from a surface and detecting the absorbtions due to the chemical bonds in any overlayers present. The resulting technique, with the acronym RAIRS (reflection-absorption infra-red spectroscopy) is a relatively rapid and non-destructive method of thin-film and surface analysis that complements the more usual UHV spectroscopies.

The maximum signal intensity that can be obtained in the RAIRS experiment depends on the type of substrate and the nature and thickness of the surface layer to be analysed (Greenler, 1966). For maximum sensitivity with a metallic substrate, an angle of incidence of around 85° to the surface normal is required. For best results, the measurements should be carried out on a Fourier Transform spectrometer with a liquid nitrogen cooled detector. Under favourable circumstances, detection limits of 0.1 monolayers can be achieved, although this does depend on the strength of the target molecule as an IR absorber.

RAIRS spectra show absorptions which obey the so-called surface selection rule, namely that only the component of the incident light with its plane of polarization normal to the sample surface can interact with bonds in the adsorbate, and consequently only those molecular vibrations with a finite component perpendicular to the specimen surface can be observed. This implies that, in principle, it is possible to obtain information on the orientation of adsorbed species on metal surfaces by comparing a bulk IR spectrum with the RAIRS spectrum of the same molecule adsorbed on the surface of interest, noting the absence of any particular absorption bands in the RAIRS spectrum.

As RAIRS is carried out with a glancing incidence geometry, it is difficult to achieve good spatial resolution. Under normal circumstances, the signal is collected from an area several millimetres square, for data acquisition times of a few minutes. Microscope objectives are available which allow an IR spectrum to be acquired from areas smaller than 50 μm in diameter, but these cannot normally operate at glancing incidence and hence they provide a means of thin-film rather than surface analysis. Recently, grazing-angle microscope accessories have become available, but they are as yet relatively untried in routine analytical practice.

Comparisons of RAIRS and XPS spectra can be very helpful in elucidating the chemical nature of an overlayer, as each technique often only gives a partial picture of the surface chemistry. For example, XPS gives no inform-ation on the degree of saturation of hydrocarbon chains whereas RAIRS is sensitive to the type of hydrocarbon bonding present. However, unlike XPS, RAIRS would not normally be used to obtain a quantitative estimate of the amount of a particular species present, or the thickness of an overlayer. In many respects, RAIRS is complementary to surface analysis by high-resolution electron energy loss spectroscopy (HREELS – see, for example, Riviere, 1990). It is a flexible technique which has been exploited in UHV experiments as a fine probe of chemical changes which occur when well-characterized surfaces are exposed to simple gas molecules, and which can also be used at atmospheric pressure as a simple and relatively rapid tool to identify the presence of particular chemical species on a surface or thin film sample.

RAIRS is only suited to the analysis of flat, reflective sample surfaces. Often samples are far from this ideal, and present rough, highly re-entrant surfaces for analysis. Supported catalysts are a typical example. In such cases a surface analysis can still be obtained using IR spectroscopy by the technique of diffuse-reflection infra-red Fourier transform spectroscopy (DRIFTS). In this technique, the signal is obtained using a mirror of large solid angle to collect diffuse rather than specular reflections and, after certain corrections have been applied, information on surface composition and chemistry can be deduced. As the name implies, a Fourier-transform spectrometer is used to give the necessary sensitivity. Both DRIFTS and RAIRS can be added to existing IR spectrometers at relatively low cost, providing an entry into surface analysis for an expenditure of perhaps three orders of magnitude less than that required for a UHV AES or XPS instrument (provided the basic IR spectrometer is already available). This in itself is a strong point in their favour. Although obviously they are incapable of replacing AES or XPS, their use should be considered in cases where the data obtained would complement the quantitative information given by the electron spectroscopies.

REFERENCES

* Selected references abstracted

J.M. Adams, R.G. Pritchard, P.I. Reid, J.M. Thomas and M.J. Walters, Anal. Chem., **49**, 2001 (1977).

K. Akaishi, A. Kizahara, Z. Kabeya, M. Kozimo and T. Gotoh, in Proc. 7th Int. Vac. Congr. 3rd Int. Conf. Solid. Surf., Eds R. Dobrezemsky, R. Rudenauer, F.P. Viebok and A. Breth, p 1477, Vienna (1977).

H.H Anderson, Appl. Phys., **9**, 56 (1979).

M.T. Anthony and M.P. Seah, J. Elec. Spectrosc. Relat. Phenom., **32**, 73 (1983).

* M.T. Anthony and M.P. Seah, Surf. Interface Anal. **6** 94 (1984).

* M.T. Anthony and M.P. Seah, Surf. Interface Anal. **6** 107 (1984).

J.C. Ashley and C.J. Tung, Surf. Interface Anal., **4**, 52 (1982).

* P. Auger, J. Phys. Radium, **6**, 205 (1925).

C.D. Bain and G.M. Whitesides, J. Phys. Chem., **93**, 1670 (1989).

T.L. Barr in "Practical Surface Analysis by Auger and X-ray Photoelectron Spectroscopy", Ed. D. Briggs and M.P. Seah, Chap. 8, p322, Wiley, Chichester (1983).

* I.R. Barkshire, M. Prutton and D.K. Skinner, Surf. Interface Anal., **17**, 213 (1991).

G. Barth, R. Linder and C.E. Bryson, Surf. Interaface Anal., **11**, 307 (1988).

O.A. Baschenko and V.I. Nefedov, J. Elec. Spectrosc. Relat. Phenom., **53**, 1 (1990).

* O.A. Baschenko and A.E. Nesmeev, J. Elec. Spectrosc. Relat. Phenom., **57**, 33 (1991).

O.A. Baschenko, F. Bokman, O. Bohman and H.O.G. Siegbahn, J. Elec. Spectrosc. Relat. Phenom., **62**, 317 (1993).

* C. Battistoni, G. Mattogno and E. Paparazzo, Surf. Interface Anal., **7**, 117 (1985).

G. Beamson, H.Q. Porter and D.W. Turner, Nature, **290**, 556 (1981).

G. Beamson, D. Briggs, S.F. Davies, I.W. Fletcher, D.T. Clark, J. Howard, U. Gelius, B. Wanneburg and P. Balzer, Surf. Interface Anal, **15**, 541 (1990).

G. Beamson and D. Briggs, "High Resolution XPS of Organic Polymers: The Scienta ESCA300 Database", Wiley, Chichester (1992).

N. Beatham and A.F. Orchard, J. Elec. Spectrosc. Relat. Phenom., **9**, 129 (1976).

H. Bender, Surf. Interface Anal., **15**, 767 (1990).

A. Benninghoven, Z. Phys., **230**, 403 (1970).

P. ten Berge, V. Young and S.R. Bates, J. Vac. Sci. Technol., **A7**, 1729 (1989).

* R.W. Bernstein and J.K. Grepstad, Surf. Interface Anal., **14**, 109 (1989).

H. Bethe, Ann. Phys. (Leipzig), **5**, 325 (1930).

* H.E. Bishop, Surf. Interface Anal., **3**, 272 (1981).

* H.E. Bishop, Surf. Interface Anal., **17**, 197 (1991).

* D. Briggs and G. Beamson. Anal. Chem., **64**, 1729 (1992).

D. Briggs, A. Brown and J.C. Vickerman, "Handbook of Static Secondary Ion Mass Spectrometry", John Wiley and Sons, Chichester (1989).

D. Briggs and M.P. Seah (Eds), "Practical Surface Analysis Volume 1 - Auger and X-ray Photoelectron Spectroscopy", Second Edition, John Wiley and Sons, Chichester (1990).

D. Briggs and M.P. Seah (Eds), "Practical Surface Analysis Volume 2 - Secondary Ion and Secondary Neutral Mass Spectrometry", John Wiley and Sons, Chichester (1992).

* R. Browning, J. Vac. Sci. Technol., **A3**, 1959 (1985).

R. Browning, D.C. Peacock and M. Prutton, Appl. Surf. Sci., **22/23**, 145 (1985).

G.T. Burstein, Mater. Sci. Eng., **42**, 207 (1980).

T.D. Bussing and P.H. Holloway, J. Vac. Sci. Technol., **A3**, 1973 (1985).

P. Caceras, B. Ralph, G.C. Allen and R.K. Wild, Surf. Interface Anal., **12**, 191 (1988).

A.F. Carley and R.W. Joyner, J. Elec. Spectrosc. Relat. Phenom., **16**, 1 (1979).

J. Carrazza and V. Leon, Surf. Interface Anal., **17**, 225 (1991).

* J. Cazaux, Scanning Electron Microscopy III, 1193 (1984).

* J. Cazaux, Appl. Surface Sci., **20**, 457 (1985).

* J. Cazaux, J. Microsc., **145**, 257 (1987).

G.P. Chambers and J. Fine, in "Practical Surface Analysis, Vol. 2: Ion and Neutral Spectroscopy", 2nd Edition, Ed. D. Briggs and M.P. Seah, Appendix 4, p 705, Wiley, Chichester (1992).

I. Chorkendorff and S. Tougaard, Appl. Surf. Sci., **29**, 101 (1987).

P.H. van Cittert, Z. Physik, **69**, 298 (1931).

D.T. Clark in "Handbook of X-ray and UV Photoelectron Spectroscopy", Ed. D. Briggs, p 212, Heyden, London (1977).

L.J. Clarke, "Surface Crystallography - An Introduction to Low Energy Electron Diffraction", Wiley, Chichester (1985).

M. Clement, J.M. Sanz and J.M. Martinez-Duart, Surf. Interface Anal., **14**, 413 (1989).

P. Coxon, J. Krizek, M. Humpherson and I.R.M. Wardell, J. Elec. Spectrosc. Relat. Phenom., **52**, 821 (1990).

* A. Cros, J. Elec. Spectrosc. Relat. Phenom., **59**, 1 (1991).

Y.M. Cross and J.E. Castle, J. Elec. Spectrosc. Relat. Phenom., **22**, 53 (1981).

* P.J. Cumpson and M.P. Seah, Surf. Interface Anal., **18**, 345 (1992).

* P.J. Cumpson and M.P. Seah, Surf. Interface Anal., **18**, 361 (1992).

A. Darlinski and J. Halbritter, Surf. Interface Anal., **10**, 223 (1987).

L.E. Davis, N.C. MacDonald, P.W. Palmberg, G.E. Riach and R.E. Weber, "Handbook of Auger Electron Spectroscopy", 2nd Edition, Physical Electronics Industries Inc., Minnesota (1976).

* E. Desimoni and U. Baider Ceipidor, J. Elec. Spectrosc. Relat. Phenom., **56**, 189 (1991).

* E. Desimoni, G.I. Casella, A. Morone and A.M. Salvi, Surf. Interface Anal., **15**, 627 (1990).

A. Dilks, in "Electron Spectroscopy Theory, Techniques and Applications", Ed. C.R. Brundle and A.D. Baker, Vol. 4, pp 277 - 359, Academic Press, London (1981).

* Z.-J. Ding and R. Shimizu, Surf. Sci., **197**, (1988).

I.W. Drummond, F.J. Street, L.P. Ogden and D.J. Surman, Scanning, **13**, 149 (1991).

* V.M. Dwyer and J.M. Richards, Surf. Interface Anal., **18**, 555 (1992).

H. Ebel, M.F. Ebel and A. Jablonski, J. Elec. Spectrosc. Relat. Phenom., **35**, 155 (1985).

H. Ebel, M.F. Ebel, P. Baldauf and A. Jablonski, Surf. Interface Anal., **12**, 172 (1988).

* H. Ebel, M.F. Ebel, C. Puchberger and R. Svagera, J. Elec. Spectrosc. Relat. Phenom., **57**, 357 (1991).

* H. Ebel, M.F. Ebel, R. Svagera, E. Winklmayr and P. Varga, J. Elec. Spectrosc. Relat. Phenom., **57**, 15 (1991).

A. van Eenbergen and E. Brunix, J. Elec. Spectrosc. Relat. Phenom., **33**, 51 (1984).

M.M. El-Gomati and M. Prutton, Surface Sci., **72**, 485 (1978).

* M.M. El-Gomati, D.C. Peacock, M. Prutton and C.G. Walker, J. Microsc. **147**, 149 (1987).

S. Evans, Surf. Interface Anal., **18**, 323 (1992).

S. Evans, R.G. Pritchard and J.M. Thomas, J. Phys. C:Solid State Phys., **10**, 2483 (1977).

S. Evans, R.G. Pritchard and J.M. Thomas, J. Elec. Spectrosc. Relat. Phenom., **14**, 341 (1978).

C.S. Fadley, Progr. Surf. Sci., **16**, 3 (1984).

M.J. Fay, A. Proctor, D.P. Hoffmann and D.M. Hercules, Anal. Chem., **63**, 1058 (1991).

* A.G. Fitzgerald, P.A. Moir and B.E. Storey, J. Elec. Spectrosc. Relat. Phenom., **59**, 127 (1992).

* J.E. Fulghum and R.W. Linton, Surf. Interface. Anal., **13**, 186 (1988).

* J.E. Fulghum, Surf. Interface Anal., **20**, 161 (1993).

S.W. Gaarenstroom, J. Vac. Sci. Technol., **16**, 600 (1979).

S.W. Gaarenstroom, Appl. Surf. Sci., **7**, 7 (1981).

S.W. Gaarenstroom, J. Vac. Sci. Technol., **20**, 458 (1982).

* S.W. Gaarenstroom, Appl. Surf. Sci., **26**, 561 (1986).

U. Gelius, Physica Scripta, **9**, 133 (1974).

U. Gelius, B. Wannberg, P. Baltzer, H. Fellner-Feldegg, G. Carlsson, C.-G. Johansson, J. Larsson, P. Munger and G. Vegerfors, J. Elec. Spectroc. Relat. Phenom., **52**, 744 (1990).

R.C. Greenler, J. Chem. Phys., **44**, 310 (1966).

O.H. Griffith, Appl. Surface Sci., **26**, 265 (1986).

* M. Gryzinsky, Phys. Rev., **A138**, 336 (1965).

S.F. Gull and J. Skilling, IEE Proceedings, **131F**, 646 (1984).

N. Gurker, M.F. Ebel and H. Ebel, Surf. Interface Anal., **5**, 13 (1983).

J. Halbritter, H. Leiste, H.-J. Mathes, P. Walk and H. Winter, Sol. State. Commun., **68**, 1061 (1988).

* P.M. Hall and J.M. Morabito, Surf. Sci., **83**, 391 (1979).

* W. Hanke, H. Ebel, M.F. Ebel, A. Jablonski and K. Hirokawa, J. Elec. Spectrosc. Relat. Phenom., **40**, 241 (1986).

L.A. Harris, J. Appl. Phys., **39**, 1419 (1968).

K. Harrison and L.B. Hazell, Surf. Interface Anal., **18**, 368 (1922).

L.B. Hazell, A.A. Rizvi, I.S. Brown and S. Ainsworth, Spectrochem. Acta, **B40**, 739 (1985).

* D.W.O. Heddle, J. Phys. E: Sci. Instrum., **4**, 589 (1971).

B.L. Henke, in "X-ray Optics and X-ray Microanalysis", Ed. H.H. Pattee, V.E. Cosslett and A. Engstrom, pp 157-172, Academic Press, New York (1963).

K.F.J. Heinrich and D.E. Newbury, "Electron Probe Quantitation", Plenum, New York (1991).

P.J. Hicks, S. Daviel, B. Wallbank and J. Comer, J. Phys. **B13**, 713 (1980).

M.D. Hill, Surf. Interface Anal., **12**, 21 (1988).

S. Hofmann, Appl. Phys., **9**, 56 (1976).

S. Hofmann, Surf. Interface Anal., **2**, 148 (1980).

S. Hofmann and J. Steffen, Surf. Interface Anal., **14**, 59 (1988).

S. Hofmann and J. Steffen, Surf. Interface Anal., **14**, 59 (1989).

* S. Hofmann, J. Elec. Spectrosc. Relat. Phenom., **59**, 15 (1992).

* S. Hofmann, Appl. Surf. Sci., **70/71**, 9 (1993).

P.H. Holloway, Surface Sci., **54**, 506 (1976).

* P.H. Holloway, Appl. Surf. Sci., **26**, 550 (1986).

A.E. Hughes and C.C. Phillips, Surf. Interace Anal., **4**, 220 (1982).

* A.E. Hughes and B.A. Sexton, J. Elec. Spectrosc. Relat. Phenom., **46**, 31 (1988).

C.P. Hunt and M.P. Seah, Surf. Interface Anal., **15**, 254 (1990).

* S. Ichimura, T. Koshikawa, T Sekine, K. Goto and H. Shimizu, Surf. Interface Anal., **11**, 94 (1988).

* S. Ichimura and R. Shimizu, Surf. Sci., **112**, 386 (1981).

S. Ichimura, D. Ze-Jun and R. Shimizu, Surf. Interface Anal., **13**, 149 (1988).

G.D. Ingram and M.P. Seah, J. Phys. E: Sci. Instrum., **22**, 242 (1989).

H. Iwasaki and S. Nakamura, Surf. Sci., **57**, 779 (1976).

* A. Jablonski, Surface Sci. **188**, 164 (1987).

A. Jablonski, Surf. Interface Anal., **14**, 659 (1989).

* A. Jablonski, J. Gryko, J. Kraer and S. Tougaard, Phys. Rev. B **39**, 61 (1989).

A. Jablonski and S. Tougaard, J. Vac. Sci. Technol., **A5**, 106 (1990).

A.P. Janssen and J.A. Venables, Surface Sci., **77**, 351 (1978).

* C. Jansson, H.S. Hansen, F. Yubero and S. Tougaard, J. Elec. Spectrosc. Relat. Phenom., **60**, 301 (1992).

P.A. Jansson, R.H. Hunt and E.K. Plyler, J. Opt. Soc. Am., **58**, 1665 (1968).

* A. Jouaiti, A. Mosser, M. Romeo and S. Shindo, J. Elec. Spectrosc. Relat. Phenom., **59**, 327 (1992).

* D. Jousset and J.P. Langeron, J. Vac. Sci. Technol., **A5**, 989 (1987).

* R. Kelly, Surf. Interface Anal., **7**, 1 (1985).

J.E. Kempf and H.H. Wagner, in "Thin Film and Depth Profile Analysis", Ed.
H. Oechsner, pp 87-102, Springer, Berlin (1984).

J. Kirschner and H.W. Etzkorn, Appl. Surf. Sci., **3**, 251 (1979).

J. Kirz and H. Rarback, Rev. Sci. Instrum., **56**, 1 (1985).

E.L. Kosarev and E. Pantos, J. Phys. E: Sci. Instrum., **16**, 537 (1983).

* H.P.C.E. Kuipers, H.C.E. van Leuven and W.M. Wisser, Surf. Interface Anal.,
8, 235 (1986).

G.G. Kurbatov and V.I. Zaporozchenko, J. Elec. Spectrosc. Relat. Phenom.,
56, 353 (1991).

* C.E. Kuyatt and J.A. Simpson, Rev. Sci. Instrum., **38**, 103 (1967).

N.Q. Lam, Surf. Interface Anal., **12**, 65 (1988).

J.P. Langeron, Surf. Interface Anal., **14**, 381 (1989).

J.P. Langeron, L. Minel, J.L. Vignes, S. Bouquet, F. Pellerin, G. Lorang, P.
Ailloud and J. Le Hericy, Surf. Sci., **138**, 610 (1984).

J.J. Lander, Phys. Rev. **91**, 1382 (1953).

P. Laty, D. Seethanen and F. Degreve, Surf. Sci., **85**, 533 (1979).

K.R. Lea and I.H. Munro, "The Synchrotron Radiation Source of Daresbury
Laboratory", Daresbury Laboratory, Warrington, UK (1980).

* R.C.G. Leckey, J. Elec. Spectrosc. Relat. Phenom., **43**, 183 (1987).

U. Littmark and W.O. Hofer, Nucl. Instrum. Methods Phys. Res., **B27**, 1
(1980).

* R.-S. Li, L.-X. Tu and Y.-Z. Sun, Surf. Sci., **163**, 67 (1985).

F. Lodders, A, Goldmann, D. Rudolph, G. Schmahl and W. Braun, J. Elec.
Spectrosc. Relat. Phenom., **60**, 1 (1992).

C. Lu and A.W. Czanderna (Eds.), "Applications of Quartz Crystal
Microbalances", Elsevier, Amsterdam (1984).

P.C. McCaslin and V. Young, Scanning Microsc., **1**, 1545 (1987).

G.E. McGuire, "Auger Electron Spectroscopy Reference Manual", Plenum Press,
New York (1979).

E.R. Malinowski and D.G. Howery, "Factor Analysis in Chemistry", Wiley, New
York (1980).

C. Malitesta and T. Rotunno, Surf. Interface Anal., **17**, 251 (1991).

N. Martensson, in "Analytical Techniques for Thin Films", Ed. K.N. Tu and R.
Rosenberg, Chap. 3 p 71, Academic Press, London (1988).

G. Mattogno and G. Righini, Surf. Interface Anal., **17**, 689 (1991).

J.M. Morabito and R.K. Lewis, Anal. Chem., **45**, 869 (1973).

A. Morris, N. Jonathan, J.M. Dyke, P.D. Francis, N. Keddar and J.D. Mills, Rev. Sci. Instrum., **55**, 172 (1984).

* S.J. Mroczkowski, J. Vac. Sci. Technol., **A7**, 1529 (1989).

V.I. Nefedov, N.P. Sergushin, I.M. Band and M.B. Trzhaskovskaya, J. Electron. Spectrosc. Relat. Phenom., **2**, 383 (1973).

V.I. Nefedov, N.P. Sergushin, Y.V. Salyn, I.M. Band and M. B. Trzhaskovskaya, J. Electron. Spectrosc. Relat. Phenom., **7**, 175 (1975).

National Institute for Standards and Technology, Office of Standard Reference data, United States Department of Commerce, Gaithersburg, Maryland 20899, USA (1989).

* I. Olefjord, H.J. Mathieu and P. Marcus, Surf. Interface Anal., **15**, 681 (1990).

J. Osterwalder, M. Sagurton, P.J. Orders, C.S. Fadley, B.D. Hermsmeier and D.J. Freidman, J. Elec. Spectrosc. Relat. Phenom., **48**, 55 (1989).

P.W. Palmberg, J. Vac. Sci. Technol., **12**, 379 (1975).

* C.G. Pantano and T.E. Madey, Appl. Surf. Sci., **7**, 115 (1981).

R. Payling and J. Szajman, J. Electron Spectrosc. Relat Phenom. **43**, 37 (1987).

D.C. Peacock, Surf. Sci., **152/153**, 895 (1985).

Perkin Elmer product information model 10-410 toroidal X-ray monochromator, Perkin Elmer Physical Electronics Division, 6509 Flying Cloud Drive, MN 55344, USA (1990).

J.E. Pollard, D.J. Trevor, J.E. Reutt, Y.T. Lee and D.A. Shirley, J. Chem. Phys. **77**, 34 (1982).

C.J. Powell, J. Elec. Spectrosc. Relat. Phenom. **47**, 197 (1988).

C.J. Powell, Ultramicroscopy, **28**, 24 (1989).

* C.J. Powell, Surf. Interface Anal., **17**, 308 (1991).

* C.J. Powell, N.E. Erickson and T.E. Madey, J. Elec. Spectrosc. Relat. Phenom., **17**, 361 (1979).

* C.J. Powell, N.E. Erickson and T.E. Madey, J. Elec. Spectrosc. Relat. Phenom., **25**, 87 (1982).

* C.J. Powell and M.P. Seah, J. Vac. Sci. Technol., **A8**, 735 (1990).

* A. Proctor and P.M.A. Sherwood, Anal. Chem., **54**, 13 (1982).

* M. Prutton, C.G.H. Walker, J.C. Greenwood, P.G. Kenny, J.C. Dee, I.R. Barkshire, R. H. Roberts and M.M. El-Gomati, Surf. Interface Anal., **17**, 71 (1991).

* E.M. Purcell, Phys. Rev., **54**, 818 (1938).

* D.E. Ramaker, Appl. Surface Sci., **21**, 243 (1985).

  P.A. Redhead and E.V. Kornelsen, "The Physical Basis of Ultrahigh Vacuum",
    American Institute of Physics (1993).

* R.F. Reilman, A. Msezane and S.T. Manson, J. Elec. Spectrosc. Relat.
    Phenom., **8**, 389 (1976).

* M. Repoux, Surf. Interface Anal., **18**, 567 (1992).

* K.H. Richter, J. Elec. Spectrosc. Relat. Phenom., **60**, 127 (1992).

  J.C. Riviere, in "Practical Surface Analysis by Auger and X-ray
    Photoelectron Spectroscopy", Chapter 2, Second Edition. Ed. D. Briggs and
    M.P. Seah, Wiley, Chichester (1990).

  J.C. Riviere, "Surface Analytical Techniques", Monographs on the Physics and
    Chemistry of Materials, Oxford Science Publishers, Clarendon Press,
    Oxford (1990).

* N. Rosenberg, M. Tholomier and E. Vicario, J. Elec. Spectrosc. Relat.
    Phenom., **46**, 331 (1988).

  M.L. Rousch, T.D. Andreadis and O.F. Goktepe, Rad. Eff., **55**, 119 (1981).

  D. Roy and J.D. Carette, "Topics in Current Physics, Vol. 4, Electron
    Spectroscopy for Chemical Analysis", pp 13-58, Springer-Verlag, Berlin
    (1977).

  H.Z. Sar-el, Rev. Sci. Instrum., **38**, 1210 (1967).

* M. Sastry, D.V. Paranjape and P. Ganguly, J. Elec. Spectrosc. Relat.
    Phenom., **59**, 243 (1992).

* A. Savitzky and M.J,E. Golay, Anal. Chem., **36**, 1627 (1964).

  M.Scharli and J. Brunner, J. Elec. Spectrosc. Relat. Phenom., **31**, 323
    (1983).

  R.D. Schnell, D. Reiger and W. Steinmann, J. Phys. E: Sci. Instrum., **17**, 221
    (1984).

* J.H. Scofield, J. Elec. Spectrosc. Relat. Phenom., **8**, 129 (1976).

* M.P. Seah and W.A. Dench, Surf. Interface Anal. **1**, 2 (1979).

  M.P. Seah, Surf. Interface Anal., **2**, 222 (1980).

  M.P. Seah, Thin Sol. Films, **81**, 279 (1981).

  M.P. Seah, M.T. Anthony and W.A. Dench, J. Phys. E: Sci. Instrum., **16**, 848
    (1983).

* M.P. Seah and M.T. Anthony, Surf. Interface Anal., **6**, 242 (1984).

* M.P. Seah, M.E. Jones and M.T. Anthony, Surf. Interface Anal., **6**, 242
    (1984).

  M.P. Seah, J. Vac. Sci. Technol. **A3**, 1330 (1985).

M.P. Seah and M. Kuehlein, Surf. Sci., **150**, 273 (1985).

* M.P. Seah, Vacuum, **36**, 399 (1986).

* M.P. Seah, Surf. Interface Anal., **9**, 85 (1986).

M.P. Seah, W.A. Dench, B. Gale, and T.E. Groves, J. Phys. E: Sci. Instrum., **21**, 351 (1988).

M.P. Seah, in "Methods of Surface Analysis", Chapter 3, Ed. J.M. Walls, Cambridge University Press, Cambridge (1988).

M.P. Seah and G.C. Smith, Surf. Interface Anal., **11**, 69 (1988).

M.P. Seah, Surf. Interface Anal. **14**, 488 (1989).

M.P. Seah, J. Elec. Spectrosc. Relat. Phenom **50**, 137 (1990).

M.P. Seah and G.C. Smith, Surf. Interface Anal. **15**, 309 (1990).

M.P. Seah, in "Practical Surface Analysis. 1- Auger and X-ray Photoelectron Spectroscopy", Second Edition, Ed. D. Briggs and M.P. Seah, Chap. 5, Wiley, Chichester (1990).

M.P. Seah and G.C. Smith, Surf. Interface Anal., **15**, 751 (1990).

M.P. Seah and G.C. Smith, Surf. Interface Anal., **15**, 701 (1990).

M.P. Seah, G.C. Smith and M.T. Anthony, Surf. Interface Anal., **15**, 293 (1990).

M.P. Seah and G.C. Smith, Surf. Interface Anal., **17**, 855 (1991).

* M.P. Seah and M. Tossa, Surf. Interface Anal., **18**, 240 (1991).

* M.P. Seah, C.P. Hunt and M. Tossa, J. Elec. Spectrosc. Relat. Phenom., **61**, 149 (1993).

* M.P. Seah, C.P. Hunt, D. Sykes, S. Valeri, R. Muller and B. Lamb, J. Elec. Spectrosc. Relat. Phenom., **61**, 173 (1993).

T. Sekine, Y. Nagasawa, M. Kudoh, Y. Sakai, A.S. Parkes, J.D. Geller, A. Mogami and K. Kirata, "Handbook of Auger Electron Spectroscopy", Jeol., Tokyo (1982).

P.M.A. Sherwood, in "Practical Surface Analysis Volume 1 - Auger and X-ray Photoelectron Spectroscopy", Second Edition, Appendix 3, p 555, Ed. D. Briggs and M.P. Seah, Wiley, Chichester (1990).

R. Shimizu, Jap. J. Appl. Phys., **22**, 1631 (1983).

* R. Shimizu, S. Ichimurs and T. Ikuta, Surf. Sci., **115** 259 (1982).

Y. Shiokawa, T. Isida and Y. Hayashi, "Auger Electron Spectra Catalogue - A Data Collection of Elements", Anelva Corporation, Tokyo (1979).

D.A. Shirly, Phys. Rev., **B5**, 4707 (1972).

K. Siegbahn, C. Nordling, A. Fahlman, R. Nordberg, K. Hamrin, J. Hedman, G. Johansson, T. Bergmark, S.E. Karlsson, I. Lindgren and B. Lindberg,

"Atomic, Molecular and Solid State Structure Studies by Means of Electron Spectroscopy", Almqvist and Wiksells, Uppsala (1976).

* H. Siegbahn, S. Svensson and M. Lundholm, J. Elec. Spectrosc. Relat. Phenom., **24** 205 (1981).

* E.N. Sickafus, Phys. Rev., **B16**, 1436 (1977).

* P. Sigmund, Phys. Rev. **184**, 383 (1969).

P. Sigmund, in "Sputtering by Particle Bombardment 1", Chap. 2, Ed. R. Behrisch, Topics in Applied Physics, Springer-Verlag, Heidelberg (1983).

* P. Sigmund, Nucl. Instrum. Methods Phys. Res., **B27**, 1 (1987).

* R. Siuda, Surf. Sci., **177** L1011 (1986).

* G.C. Smith and A.K. Livesey, Surf. Interface Anal., **19**, 175 (1992).

G.C. Smith and M.P. Seah, J. Elec. Spectrosc. Relat. Phenom., **42**, 359 (1987).

G.C. Smith and M.P. Seah, Surf. Interface Anal., **12**, 105 (1988).

G.C. Smith and M.P. Seah, Surf. Interface Anal., **16**, 144 (1990).

M.A. Smith, J. Vac. Sci. Technol., **A9**, 2309 (1991).

* J.S. Solomon, Thin Solid Films, **154**, 11 (1987).

J. Steiner, Y. Termonia and J. Deltour, Anal. Chem., **44**, 1906 (1972).

G.T.K. Swami, Surf. Interface Anal., **14**, 3 (1989).

P. Swift, Surf. Interace Anal., **4**, 47 (1982).

S. Tanuma, C.J. Powell and D.R. Penn, Surf. Sci., **192**, L849 (1987).

* S. Tanuma, C.J. Powell and D.R. Penn, Surf. Interface Anal. **11**, 577 (1988).

S. Tanuma, C.J. Powell and D.R. Penn, J. Vac. Sci. Technol. **A8**, 2213 (1990).

* S. Tanuma, C.J. Powell and D.R. Penn, Surf. Interface Anal. **17**, 911 (1991a).

* S. Tanuma, C.J. Powell and D.R. Penn, Surf. Interface Anal. **17**, 927 (1991b).

S. Tanuma, T. Sekine, K. Yoshihara, R. Shimizu, T. Homma, H. Tokutaka, K. Goto, M. Uemura, D. Fujita, A. Kurokawa, S. Ichimura, C. Oshima, M. Kurahashi, M. Kudo, Y. Hashiguchi, T. Suzuki, T. Ohmura, F. Soeda, K. Tanaka, Y. Shiowakawa and T. Hayashi, Surf. Interface Anal., **15**, 466 (1990).

L.N. Thorp and E.J. Scheibner, J. Appl. Phys., **38**, 3320 (1967).

A.L. Tofterup, Surf. Sci., **167**, 70 (1986).

* A.L. Tofterup, Surf. Sci., **227**, 157 (1988).

A.L. Tofterup, Surf. Interface Anal., **14**, 482 (1989).

H. Tokutaka, K. Nishimori and J. Matsuura, Surface Sci., **186**, 339 (1987).

S. Tougaard and B. Jorgensen, Surf. Interface Anal., **7**, 17 (1985).

* S. Tougaard, Phys. Rev. B **34**, 6779 (1986).

S. Tougaard, Surf. Sci., **172**, L503 (1986).

* S. Tougaard, Surf. Interface Anal., **8**, 257 (1986).

S. Tougaard, Surf. Interface Anal., **11**, 453 (1988).

* S. Tougaard, Appl. Surf. Sci., **32**, 332 (1988).

S. Tougaard, Surf. Sci., **216**, 343 (1989).

S. Tougaard, H.S. Hansen and M. Neumann, Surf. Sci., **244**, 125 (1991).

* N.H. Turner and W.W. Lee, Appl. Surf. Sci., **25**, 345 (1986).

B.J. Tyler, D.G. Castner and B. D. Ratner, Surf. Interface Anal., **14**, 443 (1989).

* J. Vegh, Surf. Interface Anal., **18**, 545 (1992).

* J.A. Venables, D.R. Batchelor, M. Hanbucken, C.J. Harland and G.W. Jones, Phil. Trans. R. Soc. Lond. **A318**, 243 (1986).

* J.L. Vignes, F. Pellerin, G. Lorang, S. Bouquet, J. LeHericy and J.P. Langeron, Surf. Sci., **152/153**, 957 (1985).

C.D. Wagner, Anal. Chem., **44**, 1050 (1972).

C.D. Wagner, Faraday Disc. Chem. Soc., **60**, 291 (1975).

C.D. Wagner, Anal. Chem., **49**, 1282 (1977).

* C.D. Wagner, Anal. Chem., **51**, 468 (1979).

C.D. Wagner, W.M. Riggs, L.E. Davis, J.F. Moulder, G.E. Muilenberg, "Handbook of X-ray Photoelectron Spectroscopy", Perkin-Elmer Corporation, Eden Prairie (1979).

C.D. Wagner, L.E. Davis and W.M. Riggs, Surf. Interface Anal., **2**, 53 (1980).

* C.D. Wagner, L.E. Davis, M.V. Zeller, J.A. Taylor, R.H. Raymond and L.H. Gale, Surf. Interface Anal., **3**, 211 (1981).

C.D. Wagner, in "Practical Surface Analysis Volume 1 - Auger and X-ray Photoelectron Spectroscopy", Second Edition, Ed. D. Briggs and M.P. Seah, Appendix 5, p 595, Wiley, Chichester (1990).

R.A. Waldo, M.C. Militello and S.W. Gaarenstroom, Surf. Interface Anal., **20**, 111 (1993).

J.M. Walls, I.K. Brown and D.D. Hall, Appl. Surf. Sci., **15**, 93 (1983).

* R.J. Ward and B.J. Wood, Surf. Interface Anal., **18**, 679 (1992).

J.F. Watts, J.E. Castle and S.J. Ludlam, J. Mater. Sci., **21**, 2965 (1986).

R.E. Weber and W.T. Peria, J. Appl. Phys., **38**, 2425 (1967).

W.S.M. Werner, Surf. Sci., **257**, 319 (1991).

W.S.M. Werner, W.H. Gries and H. Stori, J. Vac. Sci. Technol., **A9**, 21 (1991).

* W.S.M. Werner, W.H. Gries and H. Stori, Surf. Interface Anal., **17**, 693 (1991).

W.S.M. Werner and H. Stori, Surf. Interface Anal., **19**, 83 (1992).

W.S.M. Werner and I.S. Tilinin, Surf. Sci. **268**, L319 (1992).

* W.S.M. Werner, J. Elec. Spectrosc. Relat. Phenom., **59**, 275 (1992).

* G.K. Wertheim, J. Elec. Spectrosc. Relat. Phenom., **50**, 31 (1990).

J.H. Wilson, J. Belson and O. Auciello, in "Ion Bombardment Modification of Surfaces", Eds. O. Auciello and R. Kelly, p 225, Elsevier, Amsterdam (1984).

R.G. Wilson, F.A. Stevie and C.W. Magee, "Secondary Ion Mass Spectrometry", Wiley, New York (1989).

K. Wittmaack and F. Schulz, Thin Sol. Films, **52**, 259 (1978).

D.P. Woodruff and T. Delchar, "Modern Techniques of Surface Science", Cambridge University Press, Cambridge (1986).

Y.L. Yan, M.A. Helfand and C.R. Clayton, Appl. Surf. Sci., **37**, 395 (1989).

Ding Ze-jun, R. Shimizu and S. Ichimura, Surf. Interface Anal., **10**, 253 (1987).

X.L. Zhou and J.M. White, Appl. Surf. Sci., **35**, 52 (1988-89).

A. Zalar, Surf. Interface Anal., **9**, 41 (1986).

* P.C. Zalm, Surf. Interface Anal., **11**, 1 (1988).

Anthony, M.T. and Seah, M.P.

XPS: ENERGY CALIBRATION OF ELECTRON SPECTROMETERS 1 - AN ABSOLUTE, TRACEABLE
ENERGY CALIBRATION AND THE PROVISION OF ATOMIC REFERENCE LINE ENERGIES

Atomic standards of binding energies are presented for the calibration of X-ray photoelectron spectrometers. The binding energies are measured using a
VG Scientific ESCA 3 MK II for which the photoelectron peak positions can be
established with a measurement precision standard deviation of 5 meV. The
absolute calibration of the voltage scale is established with an eccuracy of
11 ppm over the binding energy range 0 - 1250 eV, using a measurement chain
traceable to the NPL primary voltage standards. The zero of energy is set to
+/- 11 meV on the differential of the Ni Fermi edge using Mg K$\alpha_{1,2}$ radiation
and absolute binding energies for the Cu $2p_{3/2}$, Cu L_3MM, Cu 3p, Ag $3d_{5/2}$, Ag
M_4MM and Au $4f_{7/2}$ peaks, are established with errors of 0.01 to 0.02 eV. An
analysis is made of literature calibrations to assess their zero setting and
voltage scale errors. Preliminary tests have been made to show that the
calibrations may be reproduced on instruments outside NPL using a simple work
procedure.

Reprinted with permission from Surf. Interface Anal., **6**, 95 (1984), John
Wiley and Sons Ltd., Chichester.

Anthony, M.T. and Seah, M.P.

XPS: ENERGY CALIBRATION OF ELECTRON SPECTROMETERS 2 - RESULTS OF AN
INTERLABORATORY COMPARISON

An interlaboratory test of the NPL calibration values and methodology for the
energy scale of XPS instruments, given in Part 1, has been completed. The
NPL calibration values are given for the Cu 3p, Au $4f_{7/2}$, Ag $3d_{5/2}$, Cu LMM,
Cu $2p_{3/2}$, and Ag MNN peaks for Al and Mg Kα radiations to an accuracy of 20
meV. Thirteen participants reported with their values for these peaks from
seven instrument models representative of four manufacturers. These results
are analysed in terms of random errors, voltage scaling errors and zero
offset errors. The random errors are the smallest of these errors and amount
to 30 meV, showing that instrument users can assign peak positions with very
high precision. However, the voltage scaling and zero errors range up to 700

ppm and 1363 meV respectively showing that the weakest link in the commercial measurement chains is the electronics. It is not clear if this is mainly due to drift in the electronics after the manufacturer's installation or inadequate calibration at that time. A detailed analysis of the measurements confirms the linearity of each of the instrument energy scales to within the accuracy of their random errors.

Reprinted with permission from Surf. Interface Anal., **6**, 107 (1984), John Wiley and Sons Ltd., Chichester.

Auger, P.

SUR L'EFFET PHOTOÉLECTRIQUE COMPOSÉ

Un phénomène nouveau, consistant en l'émission par un même atome d'une série de photoélectrons de vitesses diverses, a été observé a l'aide de la méthode de C.T.R. Wilson. Une interprétation théorique simple en est donnée: elle suppose que l'énergie libérée par des déplacements d'électrons dans les niveaux d'un atome excité peut se porter tout entière sur un autre électron du même atome, et le chasser sous forme de rayon β. Cette énergie prendrait donc directement la forme corpusculaire sans passer par la forme d'un rayonnement électromagnétique. L'analogie avec les rayons β des corps radioactifs est indiquée, ainsi que diverse conséquences de la théorie proposée; 9 clichés montrant le phénomène sous ses divers aspects sont reproduits.

Reprinted from J. Phys. Radium, **6**, 205 (1925).

Barkshire, I.R., Prutton, M., and Skinner, D.K.

CORRECTION OF BACKSCATTERING EFFECTS IN THE QUANTIFICATION OF AUGER DEPTH PROFILES

The calculation of surface composition from Auger peak heights requires a knowledge of the matrix effects. These are modifications to the yield of Auger electrons arising from inelastic mean free paths, the Auger back-scattering factor and sample density effects. A method for approximating the Auger backscattering factor for films of any thickness and atomic number is described. The method is based upon the assumption that the contribution of backscattered primary electrons to the Auger yield will change in proportion to the change in the electron backscattering coefficient of the film with thickness. The Auger backscattering factor so calculated varies smoothly from the value corresponding to the bulk material of the film to that of the substrate. Using known generic equations for bulk backscattering coeff-icients and factors, a simple algebraic expression is obtained. Comparison with experimental data obtained from both high and low atomic number films yields good agreement for various primary beam energies and angles of incidence.

Reprinted with permission from Surf. Interface Anal., **17**, 213 (1991), John Wiley and Sons Ltd., Chichester.

Baschenko, O.A. and Nesmeev, A.E.

THE EFFECT OF ELASTIC PHOTOELECTRON SCATTERING ON DEPTH-PROFILING BY ANGULAR RESOLVED X-RAY PHOTOELECTRON SPECTROSCOPY

By means of Monte-Carlo calculation the effect of elastic photoelectron

scattering on angular resolved X-ray photoelectron spectroscopy (ARXPS) data
has been established. A method which accounts for this effect while restor-
ing concentration profiles from ARXPS data has been proposed. A number of
model examples showing the elastic scattering leads to about a 20% decrease
of the effective photoelectron mean free paths in a solid sample have been
considered.

Reprinted with permission from J. Electron Spectrosc. Relat. Phenom., **57**, 33
(1991), Elsevier Science Publishers B.V., Amsterdam.

Battistoni, C., Mattogno, G. and Paparazzo, E.

QUANTITATIVE SURFACE ANALYSIS BY XPS: A COMPARISON AMONG DIFFERENT
QUANTITATIVE APPROACHES

Atomic ratio determinations in a series of pure inorganic compounds have been
derived from XPS peak intensity measurements taken by area via the so-called
'first principles model' (FPM) and the 'elemental sensitivity factors' (ESF)
methods. The two methods show similar degrees of accuracy, with a typical
error of about +/- 10% when a set of experimental ESF is developed.

A lower accuracy, as expected, has been obtained when the literature ESF have
been applied. Some discrepancy between the two approaches are evident when
the kinetic energy separations of the XPS elemental peaks involved in the
atomic ratio calculations is large. This aspect is discussed in the light of
instrumental responses which are of primary concern for the transferability
of ESF results from one instrument to another.

Reprinted with permission from Surf. Interface Anal., **7**, 117 (1985), John
Wiley and Sons Ltd, Chichester.

Bernstein, R.W. and Grepstad, J.K.

XPS INTENSITY ANALYSIS FOR ASSESMENT OF THICKNESS AND COMPOSITION OF THIN
OVERLAYER FILMS: APPLICATION TO CHEMICALLY ETCHED GaAs(100) SURFACES

Quantification by XPS intensity analysis of thin multicomponent overlayers,
like tha native oxides forming on compound semiconductor surfaces, is
discussed in some detail. In particular, the advantage of exploiting the
attenuated emission from substrate core levels with different probing depths
is emphasized, in order to obtain a precise measure for the overlayer thickness dimension. For GaAs(100) treated with different chemical etches, estimates are obtained for the thickness of the native surface oxide formed after
exposure to air, of 6 - 8 Å. The elemental composition of the surface oxide
is close to the stoichiometric metal:oxygen ratio of the bulk oxides Ga_2O_3
and As_2O_3, in perfect agreement with the measured core-level shifts (1.4 +/-
0.2 and 3.0 +/- 0.1 eV for the Ga and As 2p levels, respectively). On a CF_4
plasma-etched sample, a surface reaction layer of gallium fluoride was found
with a composition close to GaF_2 and an estimated thickness of 23 A. This
paper also discusses how to determine the kinetic energy dependence of the
electron escape depth, expressed in terms of an exponent m ($\lambda \propto E^m$), by
working out consistent estimates for the surface oxide (fluoride) thickness
via two different approaches. Values for m are obtained in the range 0.5 -
0.6, in excellent agreement with previously reported data on electron escape
depth for semiconductors.

Reprinted with permission from Surf. Interface Anal., **14**, 109 (1989), John
Wiley and Sons Ltd., Chichester.

Bishop, H.E.

THE EFFECTS OF PHOTOELECTRON DIFFRACTION ON QUANTITATIVE X-RAY PHOTOELECTRON
SPECTROSCOPY

Although not discussed in most reviews of quantitative x-ray photoelectron
spectroscopy (XPS), diffraction of the photoelectrons leaving a single
crystal can lead to potential errors of a factor of two or more in the
calculated surface composition. A series of experimental measurements has
been made to demonstrate the magnitude of the diffraction effects. These
results are used as a basis for a discussion of the influence of crystalline
effects on quantitative XPS. If the full solid angle of modern electron
spectrometers is used, the potential errors are relatively small and can be
minimized by adopting suitable procedures. If the acceptance angle of the
analyser is restricted for angular-resolved or for small-area measurements,
the potential errors are substantial and great care is required to avoid
them.

Reprinted with permission from Surf. Interface Anal., **17**, 197 (1991), John
Wiley and Sons Ltd, Chichester.

Bishop, H.E.

PRACTICAL PEAK AREA MEASUREMENTS IN X-RAY PHOTOELECTRON SPECTROSCOPY

An assessment of the reliability of peak area measurements using the simple
linear background has been made by comparing the results obtained by a number
of workers from a set of simulated spectra mixing the Fe 2p peaks from Fe
metal and Fe_2O_3. The simulated spectra are generated from Gaussian peaks
with a step background similar to that of Shirley, but modified to reproduce
the sloping background on the low kinetic energy side of the peak. The
results show that systematice errors in the total area of the Fe 2p peak may
be as great as 40% using the workers best estimate of the background line.
An alternative approach involving the area of both the $2p_{3/2}$ and $2p_{1/2}$ peaks
reduces this error to 17%.

Reprinted with permission from Surf. Interface Anal., **3**, 272 (1981), John
Wiley and Sons Ltd, Chichester.

Briggs, D. and Beamson, G.

PRIMARY AND SECONDARY OXYGEN-INDUCED C 1s BINDING ENERGY SHIFTS IN X-RAY
PHOTOELECTRON SPECTROSCOPY OF POLYMERS

A total of 43 polymers containing only C, H, and O have been studied using an
X-ray photoelectron spectroscopy instrument capable of the highest energy
resolution available to date. The primary aim was to reinvestigate the C 1s
binding energy shifts for oxygen functionalities and to systematically
investigate secondary shifts for the first time. In addition, this large
data set has revealed the effects of vibrational fine structure on the C 1s
component line width/shape and the effects of shake-up/shake-off (in
photoemission from C=O groups) on nonstoichiometry of C 1s component
intensities.

Reprinted with permission from Anal. Chem., **64**, 1729 (1992), American
Chemical Society.

Browning, R.

NEW METHODS FOR IMAGE COLLECTION AND ANALYSIS IN SCANNING AUGER MICROSCOPY

While scanning Auger micrographs are used extensively for illustrating the
stoichiometry of complex surfaces and for indicating areas of interest for
fine point Auger spectroscopy, there are many problems in the quantification
and analysis of Auger images. These problems include multiple contrast
mechanisms and the lack of meaningful relationships with other Auger data.
Collection of multielemental Auger images allows some new approaches to image
analysis and presentation. Information about the distribution and quantity
of elemental combinations at a surface are retrievable, and particular
combinations of elements can be imaged, such as alloy phases. Results from
the precipitate hardened alloy Al-2124 illustrate multispectral Auger
imaging.

Reprinted with permission from J. Vac. Sci. Technol., **A3**, 1959 (1985),
American Vacuum Society, New York.

Cazaux, J.

DETECTION LIMITS IN AUGER ELECTRON SPECTROSCOPY

The sources of background noise in Auger electron microanalysis are analyzed
in order to evaluate the minimum detectable concentration x_m and the minimum
number of detectable atoms y_m that can be reached. The best choices of
operating conditions (the energy E_o, intensity I_o, and spot size d_o of the
incident beam, and the duration of the experiment T_e) are deduced for bulk
and thin film analysis. The main results are: (i) The choice of E_o is not
very stringent, at least when $E_o > 5E_i(A)$, where $E_i(A)$ is the ionization
energy. (ii) For a given electron dose received by the sample, x_m is
improved by the use of the largest incident spot size while y_m is improved by
the use of the finest spot size. The results also hold for other microanalytical techniques such as electron energy loss spectroscopy or electron
probe microanalysis. (iii) Chemical identification of a single atom will be
possible on samples able to tolerate very large electron doses by using
incident beams of 10 nm or less in diameter. The expected performance of a
coincidence technique first suggested by Wittry is also discussed.

Reprinted with permission from Surf. Sci., **140**, 85 (1984), Elsevier Science
Publishers B.V., Amsterdam.

Cazaux, J.

SOME CONSIDERATIONS OF THE LATERAL RESOLUTION IN AUGER ELECTRON SPECTROSCOPY

As for all microanalytical techniques, the improvement of the lateral resolution in AES is conflicting with the optimization of sensitivity, precision
of quantification and reduction of radiation damage. Nevertheless the need
to improve this resolution is obvious; it can be obtained either by a microprobe approach or a selected area approach. In SAM analysis of bulk samples,
this lateral resolution is limited to the micron range by the backscattering
effects when the most stringent criterion is applied to the edge function
obtained experimentally. A better criterion leads to the consideration of
the line spread function which is mainly governed by the probe size.

Enlightened by some preliminary results obtained with the HB501A, the use of high incident beam voltage, 100 kV, for AES is the discussed, with also the ultimate step (?) which consists in setting the sample into the pole pieces for improving lateral resolution.

In conclusion, the lateral resolution limits of core loss spectroscopy and X-ray photoelectron spectroscopy are briefly discussed to be compared to that of AES

Reprinted with permission from J. Microsc., **145**, 257 (1987), Royal Microscopical Society, Oxford.

Cazaux, J.

THE INFLUENCE OF RADIATION DAMAGE (MICROSCOPIC CAUSES) ON THE SENSITIVITY OF AUGER ELECTRON SPECTROSCOPY AND X-RAY PHOTOELECTRON SPECTROSCOPY

When the spatial resolution of a technique is increased, the dose needed to obtain a given signal has to be increased and damage effects can occur. Consequently, there are three limits for the spatial resolution: those concerning the optics, the available incident flux, and damage to the material tested. First, these three limits in X-ray photoelectron spectroscopy (XPS) and electron-induced Auger electron spectroscopy (e-AES) are analyzed and compared. Then the minimum concentration detectable by XPS and e-AES is expressed as function of the microscopic causes (ionization of atoms and molecules) of radiation damage (defined by a cross-section Q). The differences between the techniques are strongly reduced if the two are assumed to have the same spatial resolution. The intrinsic advantage of XPS over e-AES is one or two orders of magnitude only if high doses of X-rays are available. A strategy is deduced to perform Auger analysis of sensitive materials.

Reprinted with permission from Appl. Surf. Sci. **20**, 457 (1987), Elsevier Science Publishers B.V., Amsterdam.

Cros, A.

CHARGING EFFECTS IN X-RAY PHOTOELECTRON SPECTROSCOPY

The specific problems encountered in the analysis of insulating materials by X-ray photoelectron spectroscopy are reviewed. The emphasis is put on the various methods available for obtaining useful information

Reprinted with permission from J. Electron Spectrosc. Relat. Phenom., **59**, 1 (1991), Elsevier Science Publishers B.V., Amsterdam.

Cumpson, P.J. and Seah, M.P.

RANDOM UNCERTAINTIES IN AES AND XPS: I: UNCERTAINTIES IN PEAK ENERGIES, INTENSITIES AND AREAS DERIVED FROM PEAK SYNTHESIS

We develop equations for numerically evaluating random uncertainties in Auger and X-ray photoelectron spectroscopies. First, the general statistical theorems involving the chi-squared distributions are clearly stated. These

are applied to peak synthesis to determine the standard deviation uncertainties in quantities such as peak intensity. peak energy and peak width.
General methods for their incorporation in new software are discussed. In
the meantime, we suggest a new method for determining the uncertainty values
using existing software provided by instrument manufacturers. These software
packages typically give the chi-squared value for the fit of a model spectrum
to the experimental data. It is shown that (provided random errors dominate)
a change in any one adjustable parameter to its one standard deviation limit
will cause the chi-squared value for the fitting of all the other adjustable
parameters to increase by unity. This provides a method which directly gives
the standard deviation for each adjustable parameter.

Reprinted with permission from Surf. Interface Anal., **18**, 345 (1992), John
Wiley and Sons Ltd., Chichester.

Cumpson, P.J. and Seah, M.P.

RANDOM UNCERTAINTIES IN AES AND XPS: II: QUANTIFICATION USING EITHER RELATIVE
OR ABSOLUTE MEASUREMENTS

Often, peak intensity measurements are quantified using relative sensitivity
factors to give the composition of a surface in percentage terms. This
process affects the estimation of uncertainties; the uncertainty in the final
quantification depends in a complex way on a sum involving the squares of the
uncertainties in the measures for each element.

We examine quantification in detail, and apply Bayesian statistical methods
to the problem for the first time. We find that if the intensities are
quantified by an absolute method rather than a relative one, the final uncertainties involve sums of the squares of the reciprocals of the individual
uncertainties. Thus, the combination of uncertainties in quantification by
an absolute method can improve the accuracy of the final quoted composition,
whereas the combination of uncertainties in quantification by sensitivity
factors always reduces the accuracy.

Reprinted with permission from Surf. Interface Anal., **18**, 361 (1992), John
Wiley and Sons Ltd., Chichester.

Desimoni, E. and Baider Ceipidor, U.

NEWDTP 2.2: A VERSATILE SYSTEM FOR ROUTINE XPS DATA ACQUISITION AND ANALYSIS

A system for X-ray photoelectron spectroscopy data acquisition and processing
is described, offering a reasonable compromise between flexibility,
efficiency and cost. The acquisition program was carefully designed to be
user-friendly and to safely guide the operator in obtaining single or
repeated multiple-region spectra. The processing program was also designed
for fast and reliable operations and, besides all the usual routines, allows
for spectral syntheses and differences. In particular, iterative synthesis
procedures can be performed in a completely interactive envirnment: by this
approach the chemical sense of the operator drives the iteration strategy
throughout the analysis of the spectra. A few selected spectra are processed
to describe the software operations.

Reprinted with permission from J. Electron Spectrosc. Relat. Phenom., **56**, 189
(1991), Elsevier Science Publishers B.V., Amsterdam.

Desimoni, E., Casella, G.I., Morone, A. and Salvi, A.M.

XPS DETERMINATION OF OXYGEN-CONTAINING FUNCTIONAL GROUPS ON CARBON-FIBRE
SURFACES AND THE CLEANING OF THESE SURFACES

Careful curve fitting is needed when using XPS to investigate oxidized
functional groups on carbon-fibre surfaces. This is important because the
core regions often contain a number of overlapping features some of which are
of low intensity, and the C 1s region is influenced by an asymmetric graphitic feature. The fitting approach used requires that the same content of
oxygenated species must be obtained from both the C 1s and O 1s regions. The
approach is used to analyse untreated fibres before and after cleaning.

Reprinted with permission from Surf. Interface Anal., **15**, 627 (1990), John
Wiley and Sons Ltd., Chichester.

Ding, Z.-J. and Shimizu, R.

MONTE CARLO STUDY OF BACKSCATTERING AND SECONDARY ELECTRON GENERATION

A theoretical model is proposed for Monte Carlo simulation of backscattering
and secondary electron generation by keV electrons. The purpose of this
study is to derive more accurate full energy spectra N(E), appearing as a
background in Auger electron spectroscopy (AES, from the elastic peak to the
low secondaries. The model is based on the combined use of Gryzinski's
inner-shell electron excitation function and the dielectric function for
taking account of the valence electron contribution in inelastic scattering
processes. This theoretical treatment was, first, examined for Si, Cu and Au
by deriving the inelastic mean free paths (IMFP) and comparing those with the
experimental data and then a simulation was applied for those elements at a
primary energy of 3 keV. The result has provided the energy distributions of
backscattered and cascade secondary electrons, which agree very well with
experiment. Another application has been also tosimulate how the EN(E)
spectra near the elastic peak change with the coverage of Cu atoms of several
atomic layers on Si. The present approach is particularly usewful for
obtaining fine structures in N(E) spectra if required and for describing the
behaviour of slow electrons in all kinds of solids provided that the dielectric function is available for those materials.

Reprinted with permission from Surf. Sci., **197**, 539 (1988), Elsevier Science
Publishers B.V., Amsterdam.

Dwyer, V.M. and Richards, J.M.

THE DEPTH DISTRIBUTION FUNCTION IN AUGER/XPS ANALYSIS

An analysis of a recent three-exponential approximation of the depth distribution function (DDF) for quantitative Auger/XPS analysis is presented and a
cmparison with an exact DDF obtained for an analytical solution to the Boltzmann transport equation in the transport approximation. Several problems
with the approximate DDF are identified and, using the derived analytical
form, and alternative two-exponential DDF is suggested.

Reprinted with permission from Surf. Interface Anal., **18**, 555 (1992), John
Wiley and Sons Ltd., Chichester.

Ebel, H., Ebel, M.F., Puchberger, C. and Svagera, R.

ON THE ENERGY DEPENDENCE OF ATTENUATION LENGTHS IN HYDROCARBON CONTAMINATIONS

An experimental procedure for determination of the reduced thicknesses of
thin films of contamination layers is described. It becomes possible to
determine the energy dependence of the attenuation length in contamination
layers from the angular distribution of photoelectronand Auger signals from
teh substrate at different kinetic energies. Within a restricted interval of
electron energies, the attenuation length is proportional to E_{kin}^m. The
evaluation of our experimental results gave a value of m = 0.64 in the energy
range 0.4 keV < E_{kin} < 1.4 keV. From considerations of the influences of
surface roughness and finite solid angle of acceptance on measured m values
it must be expected that the true value of m is larger than 0.64. Since
these two influences on measured charateristic photoelectron signals will
occur in practice too, we propose an effective value m = 0.64 for correction
of the energy dependent attenuation in contaminations.

Reprinted with permission from J. Electron Spectrosc. Relat. Phenom., **57**, 357
(1991), Elsevier Science Publishers B.V., Amsterdam.

Ebel, H., Ebel, M.F., Svagera, R., Winklmayr, E. and Varga, P.

A COMPARISON OF TWO XPS METHODS FOR QUANTIFICATION OF CONCENTRATION PROFILES

Binary Al-Li alloys are known for well pronounced segregation at ambient
temperatures. Literature provides a good knowledge of segregation behaviour
of these alloys. An exponential concentration profile with depth can be
assumed. We compare the concentration profiles obtained from SIMS measurements with results from XPS.

SIMS measurements [M. Vonbank and P. Varga, Vak.-Tek., **37** (1988) 220.]
performed on a specimen with 9.1 at% Li after storage at room temperature for
more than 48 h gave a surface composition of 50 at% Li and a gradient of -20
at% Li nm^{-1}. We investigated the same specimen after heat treatment for 2 h
20 min at 20°C. For comparison, XPS with variable take-off angle gives 75
at% Li and -17.8 at% Li nm^{-1} assuming an exponential depth profile, 70 at% Li
and -9.9 at% Li nm^{-1} with a linear response and our new imaging XPS method
with a linear response gives 92 at% Li and -15.4 at% Li nm^{-1}. The
differences are due to: the statistical significance of the measured data (8%
relative error in case of variable take-off angle experiments and 25% for
imaging experiments) especially due to weak Li 1s signals; the necessity of a
specimen transfer in ambient atmosphere; and uncertainties in sputter depths
of about 10%.

In principle the imaging method is similar to methods in which the surface is
etched with a constant flux Ar ion beam and changes in the composition of the
surface with time are monitored using XPS.

Reprinted with permission from J. Electron Spectrosc. Relat. Phenom., **57**, 15
(1991), Elsevier Science Publishers B.V., Amsterdam.

Fitzgerald, A.G., Moir, P.A. and Storey, B.E.

APPLICATION OF COMPUTER PROGRAMS TO MATRIX CORRECTIONS IN THE SURFACE
ANALYSIS OF AMORPHOUS FILMS

The computer programs AQUA and QUAX have been developed to enable the rapid
quantification of surface analysis by AES and XPS. The compilation of

library files of basic elemental data to enable the computer programs to
perform matrix calculations automatically is discussed. The benefits of this
system of programs in investigating the effects of using different methods of
calculation of paramters such as attenuation length (AL) or backscattering
factor on the calculated matrix factor and the final composition are demonstrated. Carry through this range of calculations with the AQUA and QUAX
computer surface analysis systems is shown to speed up the process of quantitative surface analysis by AES and XPS. In this investigation matrix factors
and surface compositions have been calculated for a number of materials
including films of amorphous silicon carbide and amorphous silicon nitride
with a range of compositions. For the amorphous silicon carnbide system, AES
and XPS quantification with a range of methods for AL correction gave good
agreement provided an SiC standard was used.

Reprinted with permission from J. Electron Spectrosc. Relat. Phenom., **59**, 127
(1992), Elsevier Science Publishers B.V., Amsterdam.

Fulghum, J.E.

DETERMINATION OF OVERLAYER THICKNESS BY ANGLE-RESOLVED XPS: A COMPARISON OF
ALGORITHMS

Angle-resolved XPS data from two well-characterized samples of SiO_2 on Si
were analyzed using three different algorithms for the determination of
overlayer thickness. The relative ratio algorithm, the Heaviside step
function approximation to the Laplace transform algorithm and the absolute
ratio algorithm were conpared in terms of the d/λ values calculated,
correlated parameters, experimental requirements, goodness-of-fit and
sensitivity to overlayer thickness variation. Although an inverse
correlation exists between the normalization parameter and d/λ for the
relative ratio algorithm, this algorithm is recommended over the absolute
ratio algorithm and the Laplace transform algorithm for the determination of
overlayer thicknesses. The relative ratio algorithm results in the most
accurate overlayer thickness ratio and is more sensitive to overlayer
thickness variations than the other algorithms.

Reprinted with permission from Surf. Interface Anal., **20**, 161 (1993), John
Wileys and Sons Ltd., Chichester.

Fulghum, J.E. and Linton, R.W.

QUANTITATION OF COVERAGES ON ROUGH SURFACES BY XPS: AN OVERVIEW

Selected methods for the XPS quantitation of surface coverages are summarized
with an emphasis on non-planar samples with overlayers, e.g. powders of high
surface area. Equations have been modified for uniform terminology and are
presented based on a photoelectron spectrometer with an E^{-1} transmission
function. Approaches which require a minimum of information from independent
analytical techniques are given special consideration. Quantitative schemes
discussed include calibration curves, surface sensitivity factors, exponential atteuation of bulk signals by overlayers, catalyst models, and the use
of subshell twins or peak shape distortions related to photoelectron energy
losses.

Reprinted with permission from Surf. Interface Anal., **13**, 186 (1988), John
Wiley and Sons Ltd., Chichester.

Gaarenstroom, S.W.

APPLICATION OF FACTOR ANALYSIS TO ELEMENTAL DETECTION LIMITS IN SPUTTER DEPTH
PROFILING

Factor analysis of Auger spectra acquired during sputter depth profiling is
superior to the conventional peak-to-peak amplitude method for determining
elemental compositions, especially when the Auger signal strength is near the
detection limit. The reason for the improvement is that factor analysis
utilizes information from all the data channels in the Auger spectrum while
the peak-to-peak amplitude method uses information from only two data
channels. In addition, factor analysis can separate much of the spectral
noise from the signal during processing, while peak-to-peak amplitude addit-
ively measures signal plus the range of spectral noise. Procedures can also
easily account for interfering species, even when their spectra are not
known. In one example, at least a factor of five improvement in the minimum
detection limit was achieved. In applcation to secondary ion mass spectrometry, only a marginal improvement in detection and precision was achieved.
This is because our existing procedure (peak area measurement) already
utilizes spectral information content fairly efficiently. However, factor
analysis is capable of handling spectral interferences that the peak area
method cannot.

Reprinted with permission from Appl. Surf. Sci., **26**, 561 (1986), Elsevier
Science Publishers B.V., Amsterdam.

El-Gomati, M.M., Peacock, D.C., Prutton, M. and Walker, C.G.

SCATTER DIAGRAMS IN ENERGY ANALYSED DIGITAL IMAGING: APPLICATION TO SCANNING
AUGER MICROSCOPY

In order to improve methods for the systematic characterization of inhomogeneous materials the procedures of multi-spectral imaging and scatter
diagram construction have been deployed. Although the techniques described
are relevant to all instruments which detect sgnals on different information
channels (e.g wavelengthe or energy analysed optical, X-ray or electron
imaging), they are illustrated with scanning Auger microscopy (SAM). The
construction and use of bivariate correlation diagrams is described by reference to simple samples consisting of patterns of Al film upon Si substrates.
The method is then applied to $LaNi_5$ and GaInAs samples each with different
signal/noise ratios and chemical characteristics. The software windowing of
scatter diagrams by computer combined with presentation of false colour
images is demonstrated. The multi-spectral Auger mapping (MULSAM) procedure
thus evolved is demonstrated to be a powerful, general analytical technique
for characterizing the number, abundance and chemistry of each type of region
in the surface of an inhomogeneous solid.

Reprinted with permission from J. Microsc. **147**, 149 (1987), Royal
Microscopical Society, Oxford.

Gryzinski, M.

CLASSICAL THEORY OF ATOMIC COLLISIONS. I. THEORY OF INELASTIC COLLISIONS

In this paper, a classical theory of inelastic atomic collisions is evolved
on the basis of the relations for binary collisions as well as for the
Coulomb collisions derived in the laboratory system of coordinates. Built up
as an approximation based on the binary collisions, i.e., the independent
pair interactions of the individual elements of the colliding systems, the
theory, with its immense simplicity, not only permits a clear qualitative
interpretation of the atomic collisions, but also describes well their
quantitative aspect. In terms of that theory, a majority of basic inelastic

processes accompanying the atomic collisions are analyzed. In particular, calculations are made for the following; (i) ionization of atoms and molecules by light particles (electrons), as well as by heavy particles (protons, deuterons), including inner-shell ionization and double ionization; (ii) excitation of single and triplet lines (excitation with exchange and without exchange); (iii) capture of electrons in orbit; (iv) slowing down of heavy charged particles (with consideration of the capture process as well as of the interaction with the Coulomb field of the nucleus); (v) inelastic scattering of electrons on atoms and molecules. According to the theory developed, the "diffraction" of elementary particles on atomic systems is explained on the basis of corpuscular mechanics; it is shown that the discrete energy states of the scatterer electron - and the anisotropy in the space orientation of their velocities in the case of crystals - are responsible for the main features of the diffraction pattern. Having at our disposal a simple theory without any arbitrary parameters except those describing the target system, we find it a useful tool for the investigation of atomic structures.

Reprinted with permission from Phys. Rev. **138**, A336 (1965), American Physical, Society.

Hall, P.M. and Morabito, J.M.

MATRIX EFFECTS IN QUANTITATIVE AUGER ANALYSIS OF DILUTE ALLOYS

In quantitative Auger analysis, the sensitivity factor of element i relative to element j is commonly set equal to the ratio of a peak height measured on a specimen of pure i to a peak height measured on a specimen of pure j. This neglects the matrix dependence of the Auger process, caused by changes in peak shape, sputtering, atomic density, escape depth, and backscattering. This paper evaluates the corresction factors required for the last three of these. For the escape depth, both Penn's analysis and the equations of Seah and Dench are evaluated. For backscattering, Reuter's relation is used. The result is two tables, each with 4860 correction factors for all combinations of the more intense transitions of fifty-five elements evaluated when the matrix is almost purely one element. The distribution of these factors has a median of 1.0 and a standard deviation of 0.5. Indivdual entries in the tables may vary by as much as a factor of three, depending on which equation is used for the escape depth. These correction factors are typically not strongly matrix dependent. Over half of them change by less than 5% in going from pure i to pure j. As more accurate data on the escape depth and backscatter corrections becomes available these tables can be evaluated for their accuracy and applicability to quantitative Auger analysis.

Reprinted with permission from Surf. Sci., **83**, 391 (1979), Elsevier Science Publishers B.V., Amsterdam.

Hanke, W., Ebel, H., Ebel, M.F., Jablonski, A. and Hirokawa, K.

QUANTITATIVE XPS - MULTILINE APPROACH

The algorithm for quantitative XPS analysis developed some years ago has been extended by a new presentation of the fundamental parameters and by using as many lines as possible from each element under investigation. It now gives the opportunity to reduce the influences of statistical (from measurement) and systematic errors (from fundamental parameters) on quantitative analysis results.

Reprinted with permission from J. Electron Spectrosc. Relat. Phenom., **40**, 241 (1986), Elsevier Science Publishers B.V., Amsterdam.

Heddle, D.W.O.

A COMPARISON OF THE ETENDUE OF ELECTRON SPECTROMETERS

The etendue, defined as the product of entrance area and solid angle, of
three dispersive electron spectrometers is calculated in terms of the
resolving power and a dimension. The cylindrical mirror used to image a
source on the surface of the inner cylinder on to the same surface has the
largest etendue. The effect of a focusing retarding field at the input of
the directionally sensitive hemispherical spectrometer is shown to increase
the etendue to such an extent that it may exceed that of the cylindrical
mirror. A similar expression is obtained for the retarding spherical grid
spectrometer.

Reprinted with permission from J. Phys. E. Sci. Instrum., **4**, 589 (1971),
Institute of Physics Publishing, Bristol.

Hofmann, S.

CHARGING AND CHARGE COMPENSATION IN AES ANALYSIS OF INSULATORS

Practical AES analysis of insulating ceramic materials is often impeded by
charging of the surface. After an outline of the basic principles of
charging and its effect on qualitative and quantitative AES, generally
applicable methods for charge compensation are described, such as reducing
the primary current density, minimizing the adsorbed current path, and
increasing the total secondary emission. Reduction of charging by generation
of a positive surface charge using additional low energy ion and/or electron
irradiation as well as deposition of a small amount of metal on the surface
are discussed. Furthermore, the effects of electron beam induced desorption,
its compensation in sputter profiling and of beam heating are estimated with
respect to quantitative AES of ceramics.

Reprinted with permission from J. Electron Spectrosc. Relat. Phenom., **59**, 15
(1992), Elsevier Science Publishers B.V., Amsterdam.

Hofmann, S.

APPROACHING THE LIMITS OF HIGH RESOLUTION DEPTH PROFILING

Depth profiling by ion sputtering in combination with surface analysis
methods such as AES, XPS, ISS and SIMS is commomnly applied to characterize
thin films and interfaces with high depth resolution. Frequently, degradation of the depth resolution is observed due to changes of surface topography and composition of the sample during sputtering. Recent progress in
sputter profiling of bilayers, multilayers and delta marker profiles of
different materials and with different methods has revealed a more detailed
knowledge about the limiting factors which impede the attainment of high
resolution. In many cases, ion bombardment induced surface topography has
been shown to be considerably reduced when using sample rotation during depth
profiling. The depth resolution is mainly governed by the altered layer
generated during sputtering and depends on ion beam characteristics (ion
mass, ion energy and angle of incidence) as well as on thermochemical and
atomic transport properties of the sample components. Under favourable
conditions, the physical limits of depth resolution of the order of a few
atomic monolayers can be approached. Comparison of theoretical modelling of
depth profiles with experimental data in many cases allows an estimation of
the resolution function and enables deconvolution of the measured profile.
The predictability of the depth resolution function for given materials and

experimental conditions mainly depends on a detailed understanding of the
physico-chemical processes during ion sputtering.

Reprinted with permission from Appl. Surf. Sci., **70/71**, 9 (1993), Elsevier
Science Publishers B.V., Amsterdam.

Holloway, P.H.

PROGRESS IN INSTRUMENTATION, DATA REDUCTION, AND DEPTH PROFILES IN AUGER
ELECTRON SPECTROSCOPY

Progress in the use of Auger elecctron spectroscopy is discussed. Specifically limits to spatial resolution in the scanning Auger microprobe from
backscattered electrons is illustrated. Determination of the chemical state
by peak shapes and energies are discussed along with the use of correction
factors in quantitative Auger electron data. Finally, the use of inverse
Laplace transforms of angle-resolved electron spectroscopy data to generate
de[pth profiles of sputtered GaAs is illustrated. It is concluded that
Gibbsian surface segregation during sputtering causes depletion of As near
the surface of GaAs.

Reprinted with permission from Appl. Surf. Sci. **26**, 550 (1986), Elsevier
Science Publishers B.V., Amsterdam.

Hughes, A.E. and Sexton, B.A.

CURVE FITTING XPS SPECTRA

This paper presents an algorithm for curve fitting XPS spectra by the damped
non-linear least squares (DNLLS) technique. DNLLS is used extensively in
optical design programs but has not previously been described in the XPS
literature. Non-linear least squares (NLLS), which is a special case of
dnlls, has been described in teh literature, but DNLLS is found in the
present study to be a superior method of curve fitting XPS spectra,
particulary when the fitted lineshape is complex.

Reprinted with permission from J. Electron Spectrosc. Relat. Phenom., **46**, 31
(1988), Elsevier Science Publishers B.V., Amsterdam.

Ichimura, S., Koshikawa, T., Sekine, T., Goto, K. and Shimizu, H.

USE OF EN(E) SPECTRA FOR QUANTITATIVE AES ANALYSIS OF Au-Cu ALLOYS

The possibility of quantitative AES analysis using the whole EN(E) spectrum
was examined for Au-Cu alloy samples. The EN(E) spectrum was measured
under $\varphi = 0^{\circ}$ (normal) incidence of 10 keV electrons for the Au-44 at% Cu
sample, and under $\varphi = 0^{\circ}$, 30°, and 45° incidence of 10 keV primary electrons
for the Au-20 at% Cu sample. Those spectra were compared with spectra which
were obtained by linear combination of pure Au and Cu spectra measured under
the same experimental conditions.

Both measured and synthesized spectra coincided very well in the energy
region lower than 1000 eV, except the region of slow secondary electrons.
The results indicate that energy distribution of backscattered electrons of
teh Au-Cu alloy can be estimated very well by linear combinations of pure
sample spectra. They also indicate that, by comparison of the shape of the
measured EN(E) spectrum with the synthesized one, the surface composition of

the Au-Cu alloy sample can be estimated with reasonable accuracy (several %
error).

Reprinted with permission from Surf. Interface Anal., **11**, 94 (1988), John
Wiley and Sons Ltd., Chichester.

Ichimura, S. and Shimizu, R.

BACKSCATTERING CORRECTION FOR QUANTITATIVE AUGER ANALYSIS 1. MONTE CARLO
CALCULATIONS OF BACKSCATTERING FACTORS FOR STANDARD MATERIALS

The electron backscattering effect which is important for the quantitative
interpretation of "matrix effects" in AES is investigated by applying teh
Monte Carlo calculation technique. The present calculation model is based on
the use of a precise elastic scattering cross-section obtained by the partial
wave expansion method, as well as the combined use of Gryzinski's excitation
function and Bethe's stopping power for inelastic scattering. Systematic
calculations of the backscattering factors were performed for over 25
materials including pure elements, compounds and alloys, which have been
widely used as standard materials for practical Auger analysis. The results
should enable the accuracy of quantitative analysis by AES to be improved.

Reprinted with permission from Surf. Sci., **112**, 386 (1981), Elsevier Science
Publisher B.V., Amsterdam.

Jablonski, A.

EFFECTS OF AUGER ELECTRON ELASTIC SCATTERING IN QUANTITATIVE AES

The Monte Carlo algorithm was developed for simulating the trajectories of
electrons elastically scattered in the solid. The distribution of scattered
angles was used to establish the influence of Auger electron elastic
collisions on the results of quantitative AES analysis. The calculations
were performed for the most pronounced KLL, L_3MM and M_5NN Auger transitions.
It turned out that due to the elastic collisions the Auger electron signal is
decreased by up to 10%. The corresponding decrease of the escape depth of
Auger electrons reaches 30% as compared with the value derived from the
inelastic mean free path. The values of the inelastic mean free path
resulting from the overlayer method may be strongly affected by elastic
scattering of Auger electrons.

Reprinted with permission from Surf. Sci., **188**, 164 (1988), Elsevier SCience
Publishers B.V., Amsterdam.

Jablonski, A., Gryko, J., Kraaer, J. and Tougaard, S.

ELASTIC ELECTRON BACKSCATTERING FROM SURFACES

Elastic electron backscattering from surfaces has been studied experimentally
and theoretically. It has been shown that the recently published P_1
approximation gives an inadequate description. Much more realistic results
are obtained from the Monte-Carlo algorithm based on differential cross
sections calculated within the partial wave expansion method. Excellent
agreement has been found between the calculated results and the experimental
data obtained in the present work or taken from the literature. The present
calculations seem to be valid for electron energies exceeding 200 eV and for
low- and medium-atomic-number elements. The possibility of measuring the

inelastic mean free path of electrons from the elastically backscattered
intensity is discussed.

Reprinted with permission from Phys. Rev. B. **39**, 61 (1989), American Physical
Society.

Jansson, C., Hansen, H.S., Yubero, F. and Tougaard, S.

ACCURACY OF THE TOUGAARD METHOD FOR QUANTITATIVE SURFACE ANALYSIS.
COMPARISON OF THE UNIVERSAL AND REELS INELASTIC CROSS SECTIONS

Three tests have been performed to investigate whether cross sections
determined experimentally from reflection electron energy loss spectroscopy
(REELS) are more accurate than the Universal cross section for background
correction of electron spectra. In the first test, the shape of the
background corrected spectra from a thin layer and from an infinitely thick
layer of the same element were compared for the Ag 3d, Au 4d, and Yb 4d
peaks. It was found that the Universal cross section is more accurate than
the experimentally determined REELS cross sections to determine consistently
the peak shape of background corrected spectra. In the second test, the
intensity ratio of the two components in the Au 4d, Cu 2p, Zn 2p, and Yb 4d
doublets were compared to the theoretical photoionization cross sections.
For both the Universal and the REELS inelastic cross sections, the obtained
intensity ratios agree well with theory. In the third test, we compare with
theory the peak intensity ratios to Au 4d of major peaks from Si, Cu, Ag, Yb,
and Au obtained by using the Universal and REELS cross sections for background correction. For metals, the two cross sections give peak intensity
ratios that are equally close to theory to within the expected uncertainties
in the theoretical peak intensity ratios. However, for Ag with application
of the Universal cross section the deviation from theory is slightly higher.
For Si the REELS cross section is clearly the most accurate.

Reprinted with permission from J. Electron Spectrosc. Relat. Phenom., **60**, 301
(1992), Elsevier Science Publishers B.V., Amsterdam.

Jouaiti, A., Mosser, A., Romeo, M. and Shindo, S.

BACKGROUND CALCULATION FOR X-RAY PHOTOELECTRON SPECTRA ANALYSIS

We develop a fitting method for the background calculation of XPS spectra in
order to determine the line shape and peak intensity of the primary emission.
We compare the spectra theoretically and experimentally, using various
methods based on the multiple inelastic scattering theory. We determine the
inelastic background parameter, i.e. the ratio of the elastic peak area to
the height of inelastic background, with which we compare the results of
theoretical calculations. We investigate the contribution of surface
excitations to the background formation using angle-resolved X-ray
photoelectron spectra.

Reprinted with permission from J. Electron Spectrosc. Relat. Phenom., **59**, 327
(1992), Elsevier Science Publishers B.V., Amsterdam.

Jousset, D. and Langeron, J.P.

ENERGY DISTRIBUTION OF PRIMARY BACKSCATTERD ELECTRONS IN AUGER ELECTRON
SPECTROSCOPY

One of the major problems for the quantification of Auger electron spectrocopy is the accurate subtraction of the background under the Auger peaks. Although many studies have been done on the energy distribution of electrons emitted by true secondary or Auger internal sources, the shape of the primary backscattered electron distribution has yet been completely reproduced only for aluminium with Monte-Carlo calculations. We develop a simple transport model where elastic and inelastic scattering are separated, following Tougaard and Sigmund [Phys. Rev. B.25, 4452 (1982)], and which gives the shape of the energy spectrum in the off-peak region with a diffusion approximation. This allows a measurement of the influence of physical parameters, like elastic and inelastic mean free paths and stopping power. Concurrently, we compare calculations with experimental spectra of copper, and propose to fit the primary induced background with an exponential law.

Reprinted with permission from J. Vac. Sci. Technol. A 5, 989 (1987), American Vacuum Society.

Kelly, R.

ON THE ROLE OF GIBBSIAN SEGREGATION IN CAUSING PREFERENTIAL SPUTTERING

It is well known that preferential sputtering with binary alloys correlates significantly with chemical binding, but only the sense and not the magnitude of the effect can be understood in these terms. It is argued here that the correlation is, in fact, an indirect one due to bombardment-induced Gibbsian (or similar) segregation. The preferentially removed component is characterized by a composition profile consisting of a one atom-layer-thick spike at the surface; this is the depth from which most sputtered atoms originate so the spike must have near-bulk composition. There is then a severely depleted subsurface region in accordance with Gibbsian segregation equilibrium (or an equivalent effect) and a final return to bulk composition. The reason for the marked correlation with chemical binding is that segregation is governed significantly more often by binding than by the alternatives of size, surface chemistry, an interstitial flux, or long-range ordering.

Reprinted with permission from Surf. Interface Anal., 7, 1, (1985), John Wiley and Sons Ltd., Chichester.

Kuipers, H.P.C.E., van Leuven, H.C.E. and Visser, W.M

THE CHARACTERIZATION OF HETEROGENEOUS CATALYSTS BY XPS BASED ON GEOMETRIC PROBABILITY 1: MONOMETALLIC CATALYSTS

Geometric probability can be used as a basis for the quantification of XPS-signals stemming from random samples such as, for instance, supported heterogeneouds catalysts. From a study of three ideal cases, namely extended layers of constant thickness, equally sized spheres and hemispheres on a randomly oriented support, it has been established that the effects of angular and layer thickness averaging are interconnected in such a way that a general model can be derived. It has been found that signal ratios as measured by XPS for convex particles are determined by the surface/volume ratios of the supported phases, nearly independent of particle shape.

Reprinted with permission from Surf. Interface Anal., 8, 235 (1986), John Wiley and Sons Ltd., Chichester.

Kuyatt, C.E. and Simpson, J.A.

ELECTRON MONOCHROMATOR DESIGN

A study has been made of all the known factors which limit the performance of
high resolution (0.07 to 0.01 eV FWHM) monochromators. These limiting
factors have been incorporated into design equations for the optimum (maximum
current output) monochromator. The conclusions are tested by performance
measurements on a prototype instrument. The results require the introduction
into the design equation of a new limiting factor, an anomalous energy spread
in dense electron beams, which is empirically determined.

Reprinted with permission from Rev. Sci. Instrum., **38**, 103 (1967), American
Institute of Physics.

Leckey, R.C.G.

RECENT DEVELOPMENTS IN ELECTRON ENERGY ANALYSERS

Recent developments in the design of electrostatic analysers for the analysis
of charged particles are reviewed. Emphasis is placed on design innovations
which permit multidetection in terms of energy and/or angle of acceptance and
on the use of position sensitive detectors in conjunction with traditional
analyser designs.

Reprinted with permission from J. Electron Spectrosc. Relat. Phenom., **43**, 183
(1987), Elsevier Science Publishers B.V., Amsterdam.

Li, R.-S., Tu. L.-X. and Sun, Y.-Z.

ENERGY DEPENDENCE OF THE SURFACE COMPOSITION CHANGES IN Au-Cu ALLOYS UNDER
Ar^+ ION BOMBARDMENT

The energy dependence of the surface composition changes in Au-Cu alloys
under Ar^+ ion bombardment has been studied. When the energy of the Ar^+ ions
is higher than 1 keV, the surface concentration of gold decreases with
bombardment time gradually approaching the steady state value, which is
somewhat lower than the bulk concentration of gold. Depth proifiling
indicates that gold is enriched at the outermost layer with a gold depleted
zone beneath it. When the energy of the Ar^+ ions is lower than 1 keV,
however, the surface concentration of gold increases with time, reaching a
maximum, then decreases toward a steady state value, which is higher than the
value of the bulk. Depth profiling indicates that gold is enriched not on;ly
in the outermost layer but alos in the subsurface region. Bombardment with
Ar^+ ions of 1 keV gives a steady state concentration of gold, which is close
to the bulk value. From these results, some possible explanation of the
difference between the present observation and an earlier result obtained by
Gillam are discussed. These results can be regarded as new evidence to
support a new point of view on preferential sputtering of binary alloys.
This suggests that the so-called preferential sputtering in most binary
alloys may be identical to the bombardment-induced Gibbsian segregation, but
when the ion energy decreases toward a near-threshold condition, the mass-
correlating preferential sputtering also plays a role.

Reprinted with permission from Surf. Sci., **163**, 67 (1985), Elsevier Science
Publishers B.V., Amsterdam.

Mroczkowski, S.J.

THE EFFECT OF ELECTRON TRANSMISSION FUNCTION ON CALCULATED AUGER SENSITIVITY FACTORS

It has been shown previously that calculated sensitivity factors can be used in some cases for the quantification of Auger elctron spectroscopy data. Internal calibration has been recommended for accurate quantification of transitions with energy < 200 eV. In this study, the effect of various electron transmission functions on calculated Auger electron spectroscopy sensitivity factors is investigated. Spectrometer resolution and its relation to intrinsic peak width are also discussed. The transmission function-corrected calculated sensitivity factors are compared to several available empirical handbook sensitivity factors. There is now improved reliability on the use of calculated values for the quantification of transitions at the lower end of the energy spectrum. When conductive, high-purity standards are routinely available, calculated and experimental sensitivity factors correlate well. There is a considerable difference when high-purity standards are not available as in the light element and lanthanide series. If accurate quantification is required and internal standards are not available, the Auger electron emission current should be correlated to the area under the peak for Auger spectra taken in the direct mode rather than peak-to-peak measurements in the derivative mode.

Reprinted with permission from J. Vac. Sci. Technol. A, **7**, 1529 (1989), American Vacuum Society.

Olefjord, I., Mathieu, H.J. and Marcus, P.

INTERCOMPARISON OF SURFACE ANALYSIS OF THIN ALUMINIUM OXIDE FILMS

A round robin test organized by the European Federation of Corrosion (EFC) Working Party on Surface Science and the Mechanisms of Corrosion and Protection, supported by the European Community Bureau of References (BCR), and further extended under VAMAS (Versailles Project on Advanced Materials and Standards), has been performed. The aim of the test was to study the reproducibility of surface analyses in different laboratories on well-documented samples by ESCA and Auger spectroscopy. The reference material is aluminium, oxidized at 250 $^\circ$C in oxygen. Test samples were distributed to 22 laboratories in Europe, North America and Japan. Data has been received from 20 laboratories.

The test results collected are: peak positions, intensities, chemical shifts. The reference energies of the ESCA spectrometers are 84.0eV for Au $4f_{7/2}$ and 932.7eV for Cu $2p_{3/2}$. The binding energies of the metallic states of Al 2p and Al 2s are 73.0 (+/- 0.1) eV and 118.0 (+/- 0.2) eV, respectively. The chemical shifts of Al^{3+} in the oxide layer are 2.8 (+/- 0.1) eV and 2.5 (+/- 0.1) eV. The oxygen content of the oxide layer, measured by nuclear reaction analysis, is 16.3 x 10^{15} atoms cm^{-2}. This corresponds to an oxide thickness, d, of 23 (+/- 1) A. The value of d/λ (λ is the attenuation length) calculated from the measured ESCA intensities is 1.13 (+/- 0.03) for Al 2p with an Mg Kα X-ray source, taking into account the take-off angle. The values of the attenuation length obtained from Al_2O_3 are $\lambda(Al^{3+})$ Mg Kα = 20.2 (+/- 2.0) Å and $\lambda(Al^{3+})$ Al Kα = 24.8 (+/- 3.3) Å. The ratios of the photoelectron yields of oxygen and aluminium are Y(O 1s)/Y(Al 2p) = 6.2 (+/- 0.8) and Y(O 1s)/Y(Al 2s) = 5.4 (+/- 0.8) with an Mg Kα X-ray source and Y(O 1s)/Y(Al 2p) = 6.6 (+/- 0.6) and Y(O 1s)/Y(Al 2s) = 4.9 (+/- 0.5) with an Al Kα X-ray source.

The LMM and KLL AES energies of Al in Al_2O_3 are 53.4 (+/- 2.8) eV and 1389.4 (+/- 3.5) eV, respectively. The corresponding values for aluminium metal are 66.4 (+/- 2.7) eV and 1394.4 (+/- 4.0) eV. For thin films where the attenuation length is of the same order of magnitude as the film thickness, an exponential fit of the experimental depth profile is proposed. The ratio of the sputtering rates of Al_2O_3 and Ta_2O_5 is thus found to be 0.72 (+/- 0.14).

The consistency between measurements performed in the different laboratories is excellent in many instances but certain problems are identified, which would deserve further work.

Reprinted with permission from Surf. Interface Anal., **15**, 681 (1990), John Wiley and Sons Ltd., Chichester.

Pantano, C.G. and Madey, T.E.

ELECTRON BEAM DAMAGE IN AUGER ELECTRON SPECTROSCOPY

The damaging effects of electronic excitation, charging and beam heating during Auger electron spectroscopy (AES) are treated. The origin, manifestation and practical consequences of the phenomena are reviewed. A damage threshold, or critical dose, for beam damage due to electronic excitation is derived and related to experimental parameters. The close correlation between the predicted thresholds and the critical doses for damage observed in typical AES analyses indicates that primary excitation processes dominate the beam damage mechanism. Charging and the electromigration of ions in glasses are also discussed in detail. It is suggested that AES analyses of materials with potential susceptibility to beam damage be executed and interpreted with caution.

Reprinted with permission from Appl. Surf. Sci., **7**, 115 (1981), Elsevier Science Publishers B.V., Amsterdam.

Powell, C.J.

FORMAL DATABASES FOR SURFACE ANALYSIS: THE CURRENT SITUATION AND FUTURE TRENDS

Evaluated data are needed for surface analysis, just as for other measurements, so that analyses can be made reliably and effectively. Until recently, the principal sources of data were handbooks of spectra (mainly of elements) and review articles with values or equations for the parameters needed. This article describes three efforts to develop formal databases that can be incorporated into computer systems. Information is first given on a recently announced X-ray photoelctron spectroscopy database, which contains over 13,000 line positions, chemical shifts and splittings, togertwher with associated software for searching by element, line type, line energy and other variables. Second, a summary is presented of a recent compilation of evaluated elemental ion sputtering yields for applications in depth profiling and secondary ion mass spectrometry. Finally, a description is given of a new database that is to contain Auger electron spectra and X-ray photoelectron spectra; the database architecture is being designed also to accommodate depth profiles and the results of measurements by other surface analysis techniques.

Reprinted with permission from Surf. Interface Anal. **17**, 308 (1991), John Wiley and Sons Ltd., Chichester.

Powell, C.J., Erickson, N.E. and Madey, I.E.

RESULTS OF A JOINT AUGER/ESCA ROUND ROBIN SPONSERED BY ASTM COMMITTEE E-42 ON SURFACE ANALYSIS. PART I. ESCA RESULTS

We report results of a round robin involving binding-energy (BE) and relative-intensity measurements on high-purity samples of gold and copper by X-ray photoelectron spectroscopy. These results were obtained on 38 different instruments manufactured by 8 companies. We found that the spread in reported BE values was typically greater than 2 eV while the spread in intensity ratios from cleaned samples was typically a factor of ten. We have analyzed the observed trends and have developed a procedure to show the contributions of systematic errors, random errors, and mistakes on BE and intensity measurements made with different individual instruments. This procedure, which leads to a plot of instrumental response as a function of electron energy, can be used by any user to compare the performance of his instrument with those reported here and to monitor performance as a function of time. At least part of the observed spreads in the reported BE and intensity data can be ascribed to erratic instrumental performance. The results of this round robin clearly demonstrate the need for improved calibration methods and operating procedures to ensure that data of known accuracy can be obtained routinely.

Reprinted with permission from J. Electron Spectrosc. Relat. Phenom., **17**, 361 (1979), Elsevier Science Publishers B.V., Amsterdam.

Powell, C.J., Erickson, N.E. and Madey, T.E.

RESULTS OF A JOINT AUGER/ESCA ROUND ROBIN SPONSORED BY ASTM COMMITTEE E-42 ON SURFACE ANALYSIS. PART II. AUGER RESULTS

We report the results of a round robin involving kinetic-energy (KE) and relative-intensity measurements on high-purity samples of copper and gold by Auger-electron spectroscopy. These results were obtained using 28 different instruments por analysers manufactured by four companies. We found that the spread in reported KE values ranged from 7 eV at a KE of 60 eV to 32 eV at a KE of 2025 eV. The total spread in reported intensity ratios ranged from a factor of 38 for the 60 eV and 920 eV peaks of Cu to a factor of 120 for the 70 eV and 2025 eV peaks of Au. We ahve analysed the observed trends in some detail. The systematic error of kinetic-energy measurements increases with kinetic energy for many instruments. Even though all instruments were adjusted with the use of 2 keV elastically scattered electrons, the s[read in the reported positions of the 2025 eV Au peak indicates that the instruments were not adequately calibrated. Examples of erratic response were found in the measurements of relative intensities; it was believed, though not proven, that the more extreme values of intensity ratios were associated with instrument malfunctions or operator mistakes. As in the similar ESCA round robin (Part I), the spread in the reported Auger kinetic energies and relative intensities clearly demonstrates the need for standards (e.g., calibration methods, operating procedures, and data analysis) to ensure that data of known accuracy can be obtained routinely. Until suitable standards are available, interested individuals may find it useful to compare measurements using their own Auger or ESCA instruments with the group results and the trends found in the round-robin results.

Reprinted from J. Electron Spectrosc. Relat. Phenom. **25**, 87 (1982), Elsevier Science Publishers B.V., Amsterdam.

Powell, C.J. and Seah, M.P.

PRECISION, ACCURACY AND UNCERTAINTY IN QUANTITATIVE SURFACE ANALYSES BY
AUGER-ELECTRON SPECTROSCOPY AND X-RAY PHOTOELECTRON SPECTROSCOPY

A quantitative surface analysis by Auger-electron spectroscopy (AES) or X-ray
photoelectron spectroscopy (XPS) requires a series of operations that typically includes instrument setup, specimen positioning, data acquisition, data
manipulation and data analysis. These operations involve a sequence of
measurements which are combined and/or compared with other data to yield an
analysis. The final result has an estimated uncertainty that reflects the
sum of the separate random and systematic errors in the various measurements
and sources of data. We identify the major steps in typical analyses and
comment on the major sources of error leading to estimates of uncertainty.
Systematic errors generally exceed those of a random nature and are associated with the ocmplex morphology of typical specimens, with parameters of
instrument performance, and with limitations of current methodology and data.
We review general measurement principles for surface analysis, the development of a suitable analytical strategy, and identify and discuss many of the
sources of error. The discussion is specific to AES and XPS but many of the
issues are relevant to other techniques of surface analysis. Finally, two
examples are presented to illustrate the sources and magnitudes of some of
the errors and the final uncertainties in some common examples of surface
analyses.

Reprinted with permission from J. Vac. Sci. Technol. **A8**, 735 (1990), American
Vacuum Society.

Proctor, A. and Sherwood, P.M.A.

DATA ANALYSIS TECHNIQUES IN X-RAY PHOTOELECTRON SPECTROSCOPY

Various data analysis techniques of importance in X-ray photoelectron
spectroscopy (XPS) are discussed. Inelastic background determination and
subtraction and the use of derivative and difference spectra are illustrated
with particular bias toward O 1s and C 1s spectra obtained from carbon
fibres. Derivative spectra, as a relatively quick method of resolution
enhancement, are shown to provide a fairly accurate, though qualitative
guide to the makeup of convoluted peak envelopes. The importance of spectral
alignment and normalization in difference spectra are illustrated, and
procedures for determining optimal results are shown for complex C 1s
spectra. Correct spectral alignment is found to be more important than
correct normalization for difference spectra.

Reprinted with permission from Anal. Chem., **54**, 13 (1982), American Chemical
Society.

Prutton, M., Walker, C.G.H., Greenwood, J.C., Kenny, P.G., Dee, J.C.,
Barkshire, I.R., Roberts, R.H. and El Gomati, M.M.

A THIRD GENERATION AUGER MICROSCOPE USING PARALLEL MULTISPECTRAL DATA
ACQUISITION AND ANALYSIS

The design and construction of an ultrahigh vacuum multi-imaging scanning
electron microscope is described. The microscope is designed to contain two
field electron emission columns and can acquire simultaneous digital images
from a 16-channel electron spectrometer, a four-quadrant back-scattered
electron (BSE) detector, an Si(Li) X-ray detector, a SEM detector and the
current flowing to ground through the sample. Because there is exact spatial

registration between corresponding pixels in each of the images, it is
possible to use the image set to make quantitative interpretations of the
surface and subsurface chemistry. This is done using mathematical
manipulations of the image set, together with models for the SEM, BSE and
Auger signals. Techniques are described for setting up the alignment and
characterizing the field of view and transmission function of the microscope
and its spectrometer. Examples of multi-imaging from simple samples are
given. The close coupling between the microscope and its control and
interpretation computers provides considerable power for the analysis of
inhomogeneous surfaces.

Reprinted with permission from Surf. Interface Anal., **17**, 71 (1991), John
Wiley and Sons Ltd., Chichester.

Purcell, E.M.

THE FOCUSSING OF CHARGED PARTICLES BY A SPHERICAL CONDENSER

The paths of charged particles traversing a portion of an ideal spherical
condenser are worked out. The section of a condenser considered is bounded
by two rays, enclosing an angle o, from the common centre of curvature, 0, of
the equipotential surfaces. It is shown that a group of particles,
homogeneous in energy, leaving a point P on a normal to one of these
boundaries and entering the condenser along this normal as a diverging
bundle, will be brought to a focus at a point Q lying on the line PO
extended, if the proper potential is applied to the condenser. This permits
the whole condenser gap to be used as a focussing energy analyzer, or
monochromator, of very large useful aperture. The velocity dispersion and
reduced velocity dispersion are calculated for the most general case, and are
found to take the same simple form as do the corresponding expressions for
the limited homogeneous magnetic field spectrograph. The expresssions for
the reduced velocity dispersionare identical in the two cases. Compensation
for edge effect is discussed. The relativistic modification of the theory
required for high speed particles is discussed and results are presented
which indicate that the simple theory of the electrostatic spectrograph may
be inadequate even for fairly low values of v/c. It is suggested that this
difficulty may be avoided by the choice of suitable instrument parameters.

An analyzer is decribed which has a useful aperture of 0.210 steradians, a
theoretical reduced dispersion of 1010, and which requires a total focussing
potential of 0.315E, where E is the particle energy in equivalent volts. The
operation of the analyser in focussing electrons accelerated by a field
designed to furnish an equivalent point source is described.

Reprinted with permission from Phys. Rev., **54**, 818 (1938), American Physical
Society.

Ramaker, D.E.

BONDING INFORMATION FROM AUGER SPECTROSCOPY

The use of Auger Spectroscopy to obtain bonding and electronic structure
information is reviewed. The methods for extracting the Auger lineshape from
the experimental spectrum are described; in particular the background subtraction and loss deconvolution techniques are presented. A prescription is
given for quantitatively interpreting the lineshape utilizating an empirically determined one-electron DOS and atomic Auger matrix elements. Final
state hole-hole correlation or localization effects on the lineshape are
emphasized. The carbon KVV lineshapes of graphite, diamond, and the carbides

are examined. Localization effects in the lineshapes are correlated with ionic bonding character.

Reprinted with permission from Appl. Surf. Sci., **21**, 243 (1985), Elsevier Science Publishers B.V., Amsterdam.

Reilman, R.F., Msezane, A. and Manson, S.T.

RELATIVE INTENSITIES IN PHOTOELECTRON SPECTROSCOPY OF ATOMS AND MOLECULES

The relative intensities of photoelectron lines is discussed. The relationship of observed intensities to angle of observation is considered as are the errors introduced by ignoring the fact that different lines may have different angular distributions. Tables of theoretical results for the angular distribution asymmetry parameter, β, are presented for incident Al K_α, Mg K_α, and Zr M_ζ radiation for all atomic ground state subshells of non-zero angular momentum. The application of these results to molecules is discussed.

Reprinted with permission from J. Electron Spectrosc. Relat. Phenom., **8**, 389 (1976), Elsevier Science Publishers B.V., Amsterdam.

Repoux, M.

COMPARISON OF BACKGROUND REMOVAL METHODS FOR XPS

Photoelectron peak areas are measured using various background removal methods. Linear, horizontal, Shirley's and Tougaard's backgrounds have been tested on many spectra for different pass energies in the spectrometer. These four values, obtained for each photelectron line, are compared together and it is shown that the first three methods give proportional results. These values are then compared to theoretical intensities for the five elements (Au, Ag, Cu, Ni, Cr) and one oxide (Al_2O_3). These theoretical calculations are based on the knowledge of the spectrometer transmission, photoionization cross-sections and inelastic mean free path, the product of whih is commonly called the relative sensitivity factor. The best agreement is obtained when Tougaard's background is removed.

Reprinted with permission from Surf. Interface Anal., **18**, 567 (1992), John Wiley and Sons Ltd., Chichester.

Richter, K.H.

DETERMINATION OF NUMERICAL DATA FROM POLYCHROMATIC ESCA SPECTRA AND THEIR VERIFICATION

The main disadvantage of commonly used pragmatic procedures for the extraction of numerical data from polychromatic ESCA spectra are discussed. It is shown that physically defined models allow a proper description of polychromatic core-level spectra, when new parameters of the excitation function (Al $K\alpha$ and Mg $K\alpha$) and a new background construction are used. The new background construction consists of a modified Tougaard background, with an additional parameter N, which accounts for no-loss regions, and a linear background. The appropriateline shapes are voigtians (atomic spectra), Doniach-Sunjic shapes convoluted with a gaussian shape (metallic samples) and experimentally determined singlet shapes, or Kuchiev-type shapes for the case of vibrationally non-resolved spectra of free and solid molecules.

The data obtained may be verified by comparison with physically well-defined
lorentzian and gaussian widths, spin-orbit splitting, etc., determined
separately or taken from the literature. A statistical analysis of the
residual spectra (here discussed for the first time) allows for the
falsification of statistically unreliable data.

Reprinted with permission from J. Electron Spectrosc. Relat. Phenom., **60**, 127
(1992), Elsevier Science Publishers B.V., Amsterdam.

Rosenberg, N., Tholomier, M. and Vicario, E.

BACKGROUND REMOVAL IN AUGER ELECTRON SPECTROSCOPY: A NEW EXPERIMENTAL
APPROACH

A review of the main available methods of background removal in Auger
electron spectroscop and X-ray photoelectron spectroscopy is given. The
major features, assumptions and results of theoretical works, which form the
basis of the present method, are presented. This method uses a convolution
technique of the experimental spectrum with the single event loss function.
It has been applied to Auger electron spectra (Si, Ag, Fe, Ni, Cu, Al).
When the Auger energy is sufficiently low (Si, Ag), it has been assumend that
Auger electrons act as a secondary electron source within a multiplet energy
range. In every case results are satisfactory.

Reprinted with permission from J. Electron Spectrosc. Relat. Phenom., **46**, 331
(1988), Elsevier Science Publishers B.V., Amsterdam.

Sastry, M., Paranjape, D.V. and Ganguly, P.

APPLICATION OF TOUGAARD BACKGROUND SUBTRACTION TO LANGMUIR-BLODGETT FILMS

The inelastic background subtraction algorithm proposed by Tougaard is
applied for the first time to the X-ray photoemission spectra of metal salts
of arachidic acid Langmuir-Blodgett films. The background deconvolution
procedure has been done using the inelastic scattering cross-section derived
from reflection electron energy loss spectroscopy and using a "universal"
inelastic scattering cross-section. Both approaches give physically
meaningful and nearly identical primary excitation spectra. The range of
validity of the deconvolution formulae is investigated by varying the
electron emission angle. Possible reasons for the applicability of the
Tougaard algorithm to materials containing molecules with dimensions of the
order of the inelastic scattering length, as are obtained for the Langmuir-
Blodgett films in this study, are discussed.

Reprinted with permission from J. Electron Spectrosc. Relat. Phenom., **59**, 243
(1992), Elsevier Science Publishers B.V., Amsterdam.

Savitzky, A. and Golay, M.J.E.

SMOOTHING AND DIFFERENTIATION OF DATA BY SIMPLIFIED LEAST SQUARES PROCEDURES

In attempting to analyze, on digital computers, data from basically
continuous physical experiments, numerical methods of performing familiar
operations must be developed. The operations of differentiation and
filtering are especially important both as an end in themselves, and as a
prelude to further treatment of the data. Numerical counterparts of analogue
devices that perform these operations, such as RC filters, are often
considered. However, the method of least squares may be used without

additional computational complexity and with coniderable improvement in the
information quality obtained. The least squares calculation may be carried
out in the computer by convolution of the data points with properly chosen
sets of integers. These sets of integers and their normalizing factors are
decribed and their use is illustrated in spectroscopic applications. The
computer programs required are relatively simple. Two examples are presented
as subroutines in the FORTRAN language.

Reprinted with permission from Anal. Chem., **36**, 1627 (1964), American
Chemical Society.

Scofield, J.H.

HARTREE-SLATER SUBSHELL PHOTOIONIZATION CROSS-SECTIONS AT 1254 AND 1487 eV

The results of calculations of photoelectric cross-sections for the $K\alpha$ lines
of magnesium at 1254 eV and of aluminium at 1487 eV are presented. All of
the subshell cross-sections are given for Z values up to 96. The calculations were carried out relativistically using the single-potential Hartree-
Slater atomic model.

Reprinted with permission from J. Electron Spectrosc. Relat. Phenom., **8**, 129
(1976), Elsevier Science Publishers B.V., Amsterdam.

Seah, M.P.

QUANTIFICATION AND MEASUREMENT BY AUGER ELECTRON SPECTROSCOPY AND X-RAY
PHOTOELECTRON SPECTROSCOPY

Of the surface analysis techniques currently in use, AES and XPS have the
greatest cumulative data and the highest quantitative accuracy. However,
many problems still exist and factors of two or more in error can occur in
published work due to use of the wrong or inappropriate equations. In this
review we shall cover the characterization of the electron spectrometer to
enable results from one instrument to be compared with those from another,
the calculations of basic contributions to the intensity to quantify the
result, and the expression of the result for various types of sample, for
instance adsorbed monolayers, surface nuclei, reaction layers, thin films,
inhomogeneous layers, etc. In the study of technological problems the
structure of the sample is often unknown and additional information from the
spectral background or the angular dependence of intensities must be
obtained. In this way, procedures may be evolved to express quantitative
results in relation to a clear model with traceable data.

Reprinted with permission from Vacuum, **36**, 399 (1986), Pergamon Journals
Ltd., Oxford.

Seah, M.P.

DATA COMPILATIONS: THEIR USE TO IMPROVE MEASUREMENT CERTAINTY IN SURFACE
ANALYSIS BY AES AND XPS

Data compilations are an integral and vital part of the development of the
measurement scheme for surface analysis. They have three main functions that
will be considered in this survey: (i) the provision of numerical values for
use in calculations of quantitative surface analysis, (ii) the provision of
unbiased data for the development of theory, and (iii) the establishment of a
benchmark to judge the value of current research. A brief review will be

made of practical data relating to AES and XPS instruments; electron guns, x-ray sources, spectrometers and detectors. Next, data compilations of experimental measurements of relative sensitivity factors, attenuation lengths and backscattering factors will be contrasted. Finally, theoretically derived data banks of cross sections, inelastic mean free paths, asymmetry parameters, etc will be considered and conclusions drawn of the need for further work.

Reprinted with permission from Surf. Interface Anal., **9**, 85 (1986), John Wiley and Sons Ltd., Chichester.

Seah, M.P. and Dench, W.A.

QUANTITATIVE ELECTRON SPECTROSCOPY OF SURFACES:A STANDARD DATA BASE FOR ELECTRON INELASTIC MEAN FREE PATHS IN SOLIDS

A compilation is presented of all published measurements of electron inelastic mean free path lengths in solids for energies in the range 0 - 10000 eV above the Fermi level. For analysis, the materials are grouped under one of the headings: element, inorganic compund, organic compound and adsorbed gas, with the path lengths each time expressed in nanometres, monolayers and milligrams per square metre. The path lengths are very high at low energies, fall to 0.1 - 0.8 nm for energies in the range 30 - 100 eV and then rise again as the energy increases further. For elements and inorganic compounds the scatter about a 'universal curve' is least when the path lengths are expressed in monolayers, λ_m. Analysis of the inter-element and inter-compound effects shows that λ_m is related to atom size and the most accurate relations are $\lambda_m = 538E^{-2} + 0.41(aE)^{\frac{1}{2}}$ for elements and $\lambda_m = 2170E^{-2} + 0.72(aE)^{\frac{1}{2}}$ for inorganic compounds, where a is the monolayer thickness (nm) and E is the electron energy above the Fermi level in eV. For organic compounds $\lambda_d = 49E^{-2} + 0.11E^{\frac{1}{2}}$ mg m^{-2}. Published general theoretical preductions for λ, valid above 150 eV, do not show as good correlations with the experimental data as the above relations.

Reprinted with permission from Surf. Interface Anal., **1**, 1 (1979), John Wiley and Sons Ltd., Chichester.

Seah, M.P. and Anthony, M.T.

QUANTITATIVE XPS: THE CALIBRATION OF SPECTROMETER INTENSITY - ENERGY RESPONSE FUNCTIONS 1 - THE ESTABLISHMENT OF REFERENCE PROCEDURES AND INSTRUMENT BEHAVIOUR

The ASTM E-42 committee round robin on XPS measurements found, in a simple survey, that the intensity-energy response of electron spectrometers was unacceptably unreliable. In this paper, new, high accuracy reference procedures are developed, to define the relative energy-intensity response function of spectrometers, for the constant E and   E/E modes, for analysts to calibrate their own instruments. Two procedures are developed, one involving the measurement of peak areas, for Cu, Ag and Au, carefully specified to reduce the effects of measurement noise, and the second, for use on well-designed spectrometers, through the measurement of the spectrum background. The measurements on Cu, Ag and Au reference foils show that intensity ratios are reproducible to within 2% standard deviation and, from preliminary tests, that the ratios of the response functions of a VG Scientific ESCALAB I and the NPL ESCA 3 Mk II is proportional to $E^{-0.094}$, except for the low pass energy of 20 eV in the ESCALAB I where the proportionality changes to $E^{-0.156}$. Similarly the ratio of the functions for a Perkin-Elmer PHI 550 and the VG Scientific ESCA 3 Mk II is proportional to $E^{-0.244}$ over the normal XPS measurement range to an accuracy of 2%.

Reprinted with permission from Surf. Interface Anal., **6**, 230 (1984), John Wiley and Sons Ltd., Chichester.

Seah, M.P., Jones, M.E. and Anthony, M.T.

QUANTITATIVE XPS: THE CALIBRATION OF SPECTROMETER INTENSITY – ENERGY RESPONSE FUNCTIONS 2 – RESULTS OF INTERLABORATORY MEASUREMENTS FOR COMMERCIAL INSTRUMENTS

An interlaboratory experiment to establish X-ray photoelectron spectrometer intensity-response functions has been completed. The measurements made by fifteen participants give results for seven instrument models from four manufacturers. Two methods of determining the response function are tested: (i) by peak area measurement and (ii) using the background data from wide scan spectra from reference foils of Cu, Ag, and Au. The peak area data show excellent reproducibility, and area ratios, when corrected from instrument response functions, are consistent to a standard deviation of 4% for Ag and Au between all instruments and conditions. The response functions relative to the VG Scientific ESCA 3 Mk II are proportional to E^n where, in the constant ΔE mode, n = 0.20 +/- 0.03 for the Leybold LHS10, –0.15 +/- 0.10 for the Perkin-Elmer PH1 550, 0.00 +/-0.003 for the VG Scientific ESCA 3, and 0.20 +/- 0.02 for the VG Scientific ESCALAB I and II. The AEI/Kratos ES 200B/300 gives a complex, but defined, function. In the constant $\Delta E/E$ mode the latter instrument gives n in the range 1.3 to 2.0 and the VG Scientific ESCALAB II 1.79 +/- 0.05.

Reprinted with permission from Surf. Interface Anal.,**6**, 242 (1984), John Wiley and Sons Ltd., Chichester.

Seah, M.P. and Tosa, M.

LINEARITY IN ELECTRON COUNTING AND DETECTION SYSTEMS

An analysis is presented, with supporting experimental measurements, of the non-linear effects in electron intensity measurements using channel electron multipliers (CEMs) in both the analogue amplification and pulse counting modes. It is shown that for Auger electron spectroscopy, where the CEMs may be used in the analogue mode, the output cuurent should not exceed 0.8r% of the CEM wall current for the linearity to be maintained within r%. In the pulse counting mode, used more generally, simple dead time corrections are possible over a limited range of counting rates. This range is defined theoretically and experimental procedures are demonstrated to show how the range may be defined, and the necessary corrections made, to retain accurate linearity within 0.5%.

Reprinted with permission from Surf. Interface Anal., **18**, 240 (1991), John Wiley and Sons Ltd., Chichester.

Seah, M.P., Hunt, C.P. and Tosa, M.

DEVELOPMENT OF A REFERENCE MATERIAL AND REFERENCE METHOD TO PROVIDE A CALIBRATION OF THE INSTRUMENT INTENSITY SCALE FOR DIFFERENTIAL AES

The irreproducibility of Auger electron spectra recorded in different laboratories arises through uncalibrated instrument functions. For spectra recorded in the direct mode a calibration method and reference materials have been developed which give the calibration for an instrument to convert spectra over the energy range 10 - 2500 eV to true spectra with a standard deviation of 3%. A reference material of solid copper and gold is developed here with reference data and a reference method to provide a similar calibration for Auger electron spectra recorded in the differential mode using a 5 keV electron beam incident at 30° to the surface normal. The calibration involves the use of both 5 and 10 eV sinusoidal analogue modulations and provides 9 calibration points between 60 and 2100 eV on the energy scale. Tests using both cylindrical mirror analyser and spherical sector instruments show that a calibration accuracy of 5% may be achieved so that all instruments analysing the same material under the same beam energy, angle of incidence, spectrometer resolution and modulation should give the same spectra within that limit.

Reprinted with permission from J. Electron Spectrosc. Relat. Phenom., **61**, 149 (1993), Elsevier Science Publishers B.V., Amsterdam.

Seah, M.P., Hunt, C.P., Sykes, D., Valeri, S., Muller, R. and Lamb, B.

INTERLABORATORY TESTS OF A COMPOSITE REFERENCE SAMPLE TO CALIBRATE AUGER ELECTRON SPECTROMETERS IN THE DIFFERENTIAL MODE

A robust reference material has been designed to calibrate the intensity/energy response function of Auger electron spectrometers operated in the differential mode. This material has been tested on six instruments from three manufacturers in five laboratories. The results show that the calibration methodology is consistent with previous work for Auger electron spectrometers used in the direct mode and that intensity variations of a factor of two in range, which would occur in quantification without the calibration, may now be reduced to around 5%.

Reprinted with permission from J. Electron Spectrosc. Relat. Phenom., **61**, 173 (1993), Elsevier Science Publishers B.V., Amsterdam

Seigbahn, H., Svensson, S. and Lundholm, M.

A NEW METHOD FOR ESCA STUDIES OF LIQUID-PHASE SAMPLES

A new technique is described for obtaining ESCA spectra from liquid phase samples. It combines temperature regulation of the liquid samples, which are continuously created as thin films on a metal backing, with the use of monochromatized Al Kα radiation for excitation. The technique is simple to implement and can be used routinely over essentially unlimited periods of recording time in the spectrometer. The results show substantially increased quality with respect to earlier measuremetns in terms of signal-to-background ratio and resolution. The technique also implies a vast increase in the number of solvents usable for future liquid-phase ESCA work.

Reprinted with permission from J. Electron Spectrosc. Relat. Phenom., **24**, 205 (1981), Elsevier Science Publishers B.V., Amsterdam.

Shimizu, R., Ichimura, S. and Ikuta, T.

BACKSCATTERING CORRECTION FOR QUANTITATIVE AUGER ANALYSIS II. VERIFICATIONS OF THE BACKSCATTERING FACTORS THROUGH QUANTIFICATION BY AES

The validity and utility of the backscattering correction factors obtained

from Monte-Carlo calculations for quantitative analysis by Auger electron
spectroscopy (AES) were examined through practical quantification of surface
compositions of binary alloys. Quantifications were attempted, first, to
access the surface composition of a sputter-deposited Ni-Pt layer, which is
probably the most appropriate test-sample with known surface composition for
surface analysis. The quantification by AES has lead to the result that the
surface composition of the layer agrees well with the bulk composition of the
sputtered Ni-Pt alloy, as expected. The composition of a sputtered Au-Cu
alloy surface was, then, examined according to the same correction procedure
as for the Ni-Pt layer, leading to the confirmation that no preferential
sputtering is observed for Au-Cu alloys by AES as Farber et al reported.

Reprinted with permission from Surf. Sci., **115**, 259 (1982), Elsevier Science
Publishers B.V., Amsterdam.

Sickafus, E.N.

LINEARIZED SECONDARY-ELECTRON CASCADES FROM THE SURFACES OF METALS. I. CLEAN
SURFACES OF HOMOGENEOUS SPECIMENS

The cascade in the secondary electron emission from atomically clean and
atomically characterized surfaces of metals has been studied in the energy
region of shortest inelastic mean free paths (10 eV < E < 1000 eV). Specimen
surfaces were cleaned in situ and characterized under ultra-high vacuum
conditions ($< 10^{-9}$ Torr) by low-energy electron diffraction and Auger
electron spectroscopy. The cascade from clean surfaces was found to consist
of linear segments when log j(E) was displayed vs. log E, where j(E) is the
emission current distribution as a function of the electron enrgy E. From
the linear characteristic and the dependence of emission on primary beam
energy E_p it is inferred that the emission current in the cascade is of the
form $j(E) = AE^{-m}E_p^{-n}$. It is shown that this functional form is comparable
with a solution of the Boltzmann diffusion equation. Deviations from
linearity are found in the form of a segmented (linear) display where
segmentation is caused by internal sources such as Auger electron sources.
The linear segments are joined near energies characteristic of bound
electrons in the sloid. These effects are used to form the basis of a new
approach to surface characterization.

Reprinted with permission from Phy. Rev. B, **16**, 1436 (1977), American
Physical Society.

Sigmund, P.

THEORY OF SPUTTERING. I. SPUTTERING YIELD OF AMORPHOUS AND POLYCHRYSTALLINE
TARGETS

Sputtering of a target by energetic ions or recoil atoms is assumed to result
from cascades of atomic collisions. The sputtering yield is calculated under
the assumption of random slowing down in an infinite medium. An integro-
differential equation for the yield is developed from the general Boltzmann
transport equation. Input quantities are the cross sections for ion-target
and target-target collisions, and atomic binding energies. Solutions of the
intergral equation are given that are asymptotically exact in the limit of
high ion energy as compared to atomic binding energies. Two main stages of
the collision cascade have to be distinguished: first, the slowing dowm of
the primary ion and all recoiling atoms that have comparable energies - these
particles determine tha spatial extent of the cascade; second, the creation
and slowing down of low-energy recoils that constitute the major part of all

the atoms set in motion. The separation between the two stages is essentially complete in the limit of high ion energy, as far as the calculation of the
sputtering yield is concerned. High-energy collisions are characterized by
Thomas-Fermi-type cross sections, while a Born-Mayer-type cross section is
applicable in the low-energy region. Electronic stopping is considered when
necessary. The separation of the cascade into two distinct stages has the
consequence that two characteristic depths are important for the qualitative
understanding of the sputtering process. First, the scattering events that
eventually lead to sputtering take place within a certain layer near the
surface, the thickness of which depends on ion mass and energy and on ion-
target geometry. In the elastic collision region, this thickness is a
sizable fraction of the ion range. Second, the majority of sputtered
particles originate from a very thin surface layer ($\sim$ 5 Å), because small
energies dominate. The general sputtering-yield formula is applied to
specific situations that are of interest for comparison with experiment.
These include backsputtering of thick targets by ion beams at perpendicular
and oblique incidence and ion energies above $\sim$100 eV, transmission sputtering
of thin foils, sputtering by recoil atoms from α-active atoms distributed
homogeneously or inhomogeneously in a thick target, sputtering of fissionable
specimens by fission fragments, and sputtering of specimens that are irradiated in the core of a reactor or bombarded with a neutron beam. There is
good agreement with experimental results on polycrystalline targets within
the estimated accuracy of the data and the input parameters entering the
theory. There is no need for adjustable parameters in the usual sense, but
specific experimental setups are discussed that allow independent checks or
accurate determination of some input quantities.

Reprinted with permission from Phys. Rev., **184**, 383 (1969), American Physical
Society.

Sigmund, P.

MECHANISMS AND THEORY OF PHYSICAL SPUTTERING BY PARTICLE IMPACT

This article summarizes theoretical work on sputtering during the past
decade, with the emphasis on elemental sputtering in the linear cascade and
spike regimes as well as alloy sputtering. The sputtering of molecules and
clusters, and electronic sputtering of insulators and biomolecular materials
are discussed more briefly, and topics like charge and excitation states of
sputtered particles as well as surface topography are left out, in view of
contemporary summary papers in this issue and elsewhere. The paper is non-
tutorial and assumes some general knowledge of the field on the part of the
reader, based e.g. on ref.[1] or ref.[2]. The discussion emphasizes
principles and methods as well as open problems rather than quantitative
results. A few general recommendations for efficient utilization of computer
simulation conclude the paper.

Reprinted with permission from Nucl. Instrum. Methods Phys. Res., **B27**, 1
(1987), Elsevier Science Publishers B.V., Amsterdam.

Siuda, R.

A PROPOSAL OF THE METHOD FOR DETERMINATION OF VERY THIN OVERLAYERS STRUCTURE
BY AES OR XPS

A new approach is proposed in the application of AES (or XPS) to
determination of the structure and total coverage of very thin overlayers of
fixed structure.

Reprinted with permission from Surf. Sci., **177**, L1011 (1986), Elsevier
Science Publishers B.V., Amsterdam.

Smith, G.C. and Livesey, A.K.

MAXIMUM ENTROPY: A NEW APPROACH TO NON-DESTRUCTIVE DECONVOLUTION OF DEPTH
PROFILES FROM ANGLE-DEPENDENT XPS

The application of the maximum entropy method to non-destructive depth
profiling by angle-dependent XPS is described. The algorithm gives the set
of depth profiles that has the maximum Skilling-Jaynes entropy, subject to
the condition that the calculated data agree with the measured data within
the experimental precision. The method does not require an inverse
transform, is robust to experimental bnoise and is not restricted to small
numbers of components. The programme can determine which of a set of prior
estimates for the depth profiles is most probable; however, the
reconstruction near the surface is virtually independent of this choice.
Further, the method also estimates the accuracy of the reconstruction froma
single data set. It is illustrated using model data and by a re-analysis of
angle-resolved XPS data sets available in the literature.

Reprinted with permission from Surf. Interface Anal., **19**, 175 (1992), John
Wiley and Sons Ltd., Chichester.

Solomon, S.J.

FACTOR ANALYSIS IN SURFACE SPECTROSCOPIES

Computational techniques to improve resolution, to remove interferences and
to indentify chemical components in recorded spectra are becoming commonplace
in surface analysis. One technique, known as factor analysis, is
particularly useful for extracting chem,ical information and removing peak
overlap interferences from data sets. Its power lies in its ability to
identify and quantify the componenets constituting a chemical or physical
system. For example, if a depth profile data set reflects changing
chemistry, factor analysis can be used (1) to identify and quantify the
chemical components within the profiled layer and (2) to reconstruct the
profile to show the depth distribution of the components.

The use of factor analysis in surface spectroscopies is reviewed in this
paper. Factor analysis has been used to enhance chemical bonding information
from thin films and interfaces studied with techniques such as Auger electron
spectroscopy, secondary ion mass spectrometry and X-ray photoelectron spectroscopy. In addition, factor analysis can be used to reconstruct elemental
profiles from Auger and ion scattering spectroscopies when peaks of interest
overlap each other during the course of a depth profile analysis.

Reprinted with permission from Thin Solid Films, **154**, 11 (1987), Elsevier
Science Publishers B.V., Amsterdam.

Tanuma, S., Powell, C.J. and Penn, D.R.

CALCULATIONS OF ELECTRON INELASTIC MEAN FREE PATHS FOR 31 MATERIALS

We present new calculations of electron inelastic mean free paths (IMFPs) for
200-2000 eV electrons in 27 elements (C, Mg, Al, Si, Ti, V, Cr, Fe, Ni, Cu,
Y, Zr, Nb, Mo, Ru, Rh, Pd, Ag, Hf, Ta, W, Re, Os, Ir, Pt, Au and Bi) and four
compounds (LiF, SiO_2, ZnS and Al_2O_3). These calculations are based on an

algorithm due to Penn which makes use of experimental optical data (to
represent the dependence of the inelastic scattering probability on energy
loss) and the theoretical Lindhard dielectric function (to represent the
dependence of the scattering probability on momentum transfer). Our
calculated IMFPs were fitted to the Bethe equation for inelastic electron
scattering in matter; the two parameters of the Bethe equation were then
empirically related to several material constants. The resulting general
IMFP formula is believed to be useful for predicting the IMFP dependence on
electron energy for a given material and the material-dependence for a given
energy. The new formula also appears to be a reasonable but more approximate
guide to electron attenuation lengths.

Reprinted with permission from Surf. Interface Anal., **11**, 577 (1988), John
Wiley and Sons Ltd., Chichester.

Tanuma, S., Powell, C.J. and Penn, D.R.

CALCULATIONS OF ELECTRON INELASTIC MEAN FREE PATHS II. DATA FOR 27 ELEMENTS
OVER THE 50-2000 eV RANGE

We report calculations of electron inelastic mean free paths (IMFPs) for 50-
2000 eV electrons in a group of 27 elements (C, Mg, Al, Si, Ti, V, Cr, Fe,
Ni, Cu, Y, Zr, Nb. Mo, Ru, Rh, Pd, Ag, Hf, Ta, W, Re, Os, Ir, Pt, Au and Bi).
This work extends our previous calculations (Surf. Interface Anal. 11, 577
(1988)) for the 200-2000 eV range. Substantial variations were found in the
shapes of the IMFP versus energy curves from element to element over the 50-
200 eV range and we attribute these variations to the different inelastic
scattering properties of each material. Our calculated IMFPs were fitted to
a modified form of the Bethe equation for inelastic electron scattering in
matter; this equation has four parameters. These four parameters could be
empirically related to several material parameters for our group of elements
(atomic weight, bulk density and number of valence electron per atom). IMFPs
were calculated from these empirical expressions and we found that the root
mean square diffence between these IMFPs and those initially calculated was
13%. The modified Bethe equation and our expressions for the four parameters
can therefore be used to estimate IMFPs in other materials. The uncertainties in the algorithm used for our IMFP calculation are difficult to estimate
but are beleived to be largely systematic. Since the same algorithm has been
used for calculating IMFPs, our predictive IMFP formula is considered to be
particularly useful for predicting the IMFP dependence on energy in the 50-
2000 eV range and the material dependence for a given energy.

Reprinted with permission from Surf. Interface Anal., **17**, 911 (1991), John
Wiley and Sons Ltd., Chichester.

Tanuma, S., Powell, C.J. and Penn, D.R.

CALCULATIONS OF ELECTRON INELASTIC MEAN FREE PATHS III. DATA FOR 15 INORGANIC
COMPOUNDS OVER THE 50-2000 eV RANGE

We report calculations of electron inelastic mean free paths (IMFPs) of 50-
20000 eV electrons in a group of 15 inorganic compounds (Al_2O_3, GaAs, GaP,
InP, InSb, KCl, LiF, NaCl, PbS, PbTe, SiC, Si_3N_4, SiO_2 and ZnS). As was
found in similar calculations for a group of 27 elements, there are
substantial differences in the shapes of the IMFP versus energy curves from
compound to compound for energies below 200 eV; these differences are
associated with the different inelastic electron scattering characteristics
of each material. Comparisons are made of the calculated IMFPs and the
values calculated from the predictive IMFP formula TPP-2 developed from the

IMFP calculations for the elements. Daviations in this comparison are found, which correlated with uncertainties of the optical data from which the IMFPs were calculated. The TPP-2 IMFP formula is therefore believed to be a more reliable means for determining IMFPs for these compounds than the direct calculations.

Reprinted with permission from Surf. Interface Anal., **17**, 927 (1991), John Wiley and Sons Ltd., Chichester.

Tofterup, A.L.

INELASTIC BACKGROUND SUBTRACTION FORMULAS AND ENERGY SPECTRA IN XPS AND AES

In order to subtract the inelastic background signal from measured energy spectra in quantitative electron spectroscopy, the inelastic scattering cross section must be knowm. Here, a realistic model cross section is utilized. Sharp primary spectra as well as Guassians and Lorentzians have been considered, and analytical expressions derived for energy spectra and inelastic background signals. Only a normalization constant and a parameter depending upon depth attenuation of the primary electron sources and upon the influence of elastic scattering enter the formulas. Comparisons with background signals deconvoluted from measured spectra were performed, and a good agreement was seen.

Reprinted with permission from Surf. Sci. **227**, 157 (1990), Elsevier Science Publishers B.V., Amsterdam.

Tougaard, S.

BACKGROUND REMOVAL IN X-RAY PHOTOELECTRON SPECTROSCOPY: RELATIVE IMPORTANCE OF INTRINSIC AND EXTRINSIC PROCESSES

The shape of all major X-ray excited peaks, including Auger transitions, from Cu, Ag, and Au are studied. The intesity contribution from inelastic (extrinsic) processes are removed. In all cases the primary excitation spectra thus determined are found to consist of a peak and a tail of (intrinsic) electrons which extends 50 eV below the peak energy. Beyond this energy essentially all measured intensity is accounted for without the use of adjustable parameters. The ratio of intrinsic to extrinsic contribution to the measured intensity is of the same order as previously found for the simple metals.

Reprinted from Phys. Rev. B, **34**, 6779 (1986), American Physical Society.

Tougaard, S.

IN-DEPTH CONCENTRATION PROFILE INFORMATION THROUGH ANALYSIS OF THE ENTIRE XPS PEAK SHAPE

A simple procedure, for non-destructive extraction of concentration depth profile information, through analysis of the ratio of the XPS peak area to the increase in background signal below the peak, was previously developed. However, since only one point in the background is used in the analysis, the detailed in-depth profile cannot be determined. In the present paper, a method is defined which takes into account the shape of the inelastic background signal in a wide energy range. With the procedure, it is possible, through a simple analysis of a single XPS spectrum, to (a) decide

whether the in-depth concentration varies roughly exponentially with depth
and (b) determine the characteristic length of the depth profile.

Reprinted with permission from Appl. Surf. Sci., **32**, 332 (1988), Elsevier
Science Publishers B.V., Amsterdam.

Tougaard, S.

QUANTITATIVE ANALYSIS OF THE INELASTIC BACKGROUND IN SURFACE ELECTRON
SPECTROSCOPY

Recent progress, in the quantitative analysis of the distortion of surface
electron spectra, resulting from inelastic electron scattering, is reviewed.

A physical model for the problem is formulated and based on this, basic
deconvolution formulae are found. In the formulae, detailed information on
the differential inelastic electron scattering cross-section K is essential.
Methods to determine K are studied and developed. One method is based on a
dielectric response calculation. Furthermore, it is found that for the noble
an transition metals, K is sufficiently similar that a simple 'universal'
cross-section can be defined. In addition, methods for experimental
determination of cross-sections are found and in particular a formula is
derived which allows the extraction of direct quantitative cross-section
information through a simple analysis of REELS spectra.

The importance of proper background correction is investigated for model
electron spectra. Experimental XPS spectra are also studied and it is found
that for both simple and noble metals, the primary excitation spectra extend
2 - 3 inelastic scattering events, i.e. $\sim$50 eV, below the peak energy.
Beyond this energy, the spectra are of zero intensity over a wide energy
range. Besides, experimental spectra are analysed on the basis of
experimentally determined cross-sections and the results are discussed, in
relation to the analysis based on theoretical cross-sections.

The importance of defining simple procedures for routine analysis of spectra
is stressed and it is demonstrated that the proposed 'universal' cross-
sedtion provides a very simple method for approximate spectral analysis.

Finally, some important issues to be considered in the near future are
pointed out.

Reprinted with permission from Surf. Interface Anal., **11**, 453 (1988), John
Wiley and Sons Ltd., Chichester

Turner, N.H. and Lee, W.W.

QUANTITATIVE ANALYSIS OF MIXTURES BY AUGER ELECTRON SPECTROSCOPY

In this study five different procedures for quantitative analysis were
investigated for the treatment of d(EN(E))/dE Auger Electron Spectroscopy
(AES) data. A mixture of $Ni(OH)_2$ and $Co(OH)_2$ was produced by
electrodeposition; variations in surface composition induced by sputtering
were used for the tests. The AES LMM spectral region of Ni and Co were
employed, since several of these lines overlap for these elements. The first
two methods tested were the measurement of peak-to-peak height and of the
negative beak-to-background intensiity. The third procedure used the linear
addition of the AES line shapes of the individual components where the best
fit is made by least-squares minimization. The fourth method em[ployed
Principal Component Analysis or Factorial Analysis. The last method

investigated was the subtraction of standard spectra. All of the methods
gave similar results except when little, if any, Co was present. From an
analysis of the different approaches, three of the procedures appeared to
offer some advantages. If only analogue data is available, then the negative
peak-to-baseline method is preferable in most cases over the peak-to-peak
height procedure. With digital data, Factorial Analysis or the subtraction
of standard spectra appear to be more general than the addition of standard
spectra, but each method has its own drawbacks. Subtraction of the
contributions of secondary and redistributed primary electrons was needed for
the latter three procedures.

Reprinted from Appl. Surf. Sci., **25**, 345 (1986), Elsevier Science Publishers
B.V., Amsterdam.

Vegh, J.

XPS PEAK SHAPES ACCOUNTING FOR THE INELASTIC SCATTERING OF PHOTOELECTRONS

A simple and fast method to account for inelastic scattering of photoelectrons is presented. A new peak shape class is introduced, which consists
of a conventional photopeak and a tail originating from the inelastic scattering of the electrons, created originally in the photopeak. The proposed
peak shapes with inelastic tail can be described using up to three additional
parameters. The method has both practical and computational advantages.

Reprinted with permission from Surf. Interface Anal., **18**, 545 (1992), John
Wiley and Sons Ltd., Chichester.

Venables, J.A., Batchelor, D.R., Hanbuecken, M., Harland, C.J. and Jones,
G.W.

SURFACE MICROSCOPY WITH SCANNED ELECTRON BEAMS

A brief review is given of the methods that are available for studying
surfaces on a microscopic scale. The use of finely focused scanned electron
beams is described in detail. Examples are given of Auger and secondary
electron spectroscopy and microscopy, and of diffraction techniques. These
examples are largely taken from recent work of the authers on Ag layers on
bulk single-crystal Si(111), Si(100) and W(110) surfaces, but applications to
other materials and to thin films are also discussed. Future developments
are briefly outlined.

Reprinted with permission from Phil. Trans. R. Soc. Lond. **A 318**, 243 (1986),
The Royal Society, London.

Vignes, J.L., Pellerin, F., Lorang, G., Bouquet, S., Le Hericy, J. and
Langeron, J.P.

DEPTH INFORMATION IN AUGER ELECTRON SPECTROSCOPY

Two different methods giving non destructive depth information in Auger
electron spectroscopy (AES) are described. (i) The first one is based on the
Auger ratio P/B (ratio of the Auger peak to the background at the same
energy). In the case of a binary alloy XY, and as soon as the Auger electron
mean free paths of the two elements are sufficiently different, it is possible to obtain a depth information by plotting $(P/B)_x$ versus $(P/B)_y$. (ii)
The second one is more general and uses the ratio of the Auger tail height
(T) to the Auger peak height (P). This ratio (T/P) is independent of the

concentration of the element. It is only dependent on its depth distribution. Both methods are applied to the experimental example of a silver
deposit on a Si(111) surface.

Reprinted with permission from Surf. Sci., **152/153**, 957 (1985), Elsevier
Science Publishers B.V., Amsterdam.

Wagner, C.D.

TWO-DIMENSIONAL CHEMICAL STATE PLOTS: A STANDARDIZED DATA SET FOR USE IN
IDENTIFYING CHEMICAL STATES BY X-RAY PHOTOELECTRON SPECTROSCOPY

The combined use of both photoelectron and X-ray excited Auger lines
increases the utility of ESCA for identifying chemical states. A useful
format for displaying reference data is the two-dimensional plot where the
kinetic energy of the sharpest Auger line is plotted vs. the binding energy
of the most intense photoelectron line. A compilation of data for 24
elements is presented, including critically evaluated data from the literature which have been referenced to a uniform calibration line. The format
of the plots is applicable to data obtained with ionizing photons of any
energy. The data included for silicon, bromine and tungsten were obtained
with higher energy X-rays from a Au X-ray source.

Reprinted with permission from Anal. Chem., **51**, 468 (1979), American Chemical
Society.

Wagner, C.D., Davis, L.E., Zeller, M.V., Taylor, J.A., Raymond, R.H. and
Gale, L.H.

EMPIRICAL ATOMIC SENSITIVITY FACTORS FOR QUANTITATIVE ANALYSIS BY ELECTRON
SPECTROSCOPY FOR CHEMICAL ANALYSIS

Quantitative information from electron spectroscopy for chemical analysis
requires the use of suitable atomic sensitivity factors. An empirical set
has been developed, based upon data from 135 compounds of 62 elements. Data
upon which the factors are based are intensity ratios of spectral lines with
F 1s as a primary standard, value unity, and K $2p_{3/2}$ as a secondary standard.
The data were obtained on two instruments, the Physical Electronics 550 and
the Varian IEE-15, two instruments that use electron retardation for scanning, with constant pass energy. The agreement in data from the two instruments on the same compounds is good. How closely the data can apply to instruments with input lens systems is not known. Calculated cross-section data
plotted against binding energy on a log-log plot provide curves composed of
simple linear segments for the strong lines: 1s, $2p_{3/2}$, $3d_{5/2}$, and $4f_{7/2}$.
Similarly, the plots for the secondary lines, 2s, $3p_{3/2}$, $4d_{5/2}$ and $5d_{5/2}$, are
shown to be composed of linear segments. Theoretical sensitivity factors
relative to F 1s should fall on similar curves, with minor correction for the
combined energy dependence of instrumental transmission and mean free path.
Experimental intensity ratios relative to F 1s were plotted similarly, and
best fit curves were calculated using the shapes of the theoretical curves as
a guide. The intercepts of these best fit curves with appropriate binding
energies provide sensitivity factors for the strong lines and the secondary
lines for all of the elements except the rare earths and the first series of
transition metals. For these elements the sensitivity factors are lower than
expected, and variable, because of multi-electron processes that vary with
chemical state. From the data it can be shown that many of the commonly-
accepted calculated cross-section data must be significantly in error - as
much as 40% in some cases for the strong lines, and far more than that for
some of the secondary lines.

Reprinted with permission from Surf. Interface Anal., 3, 211 (1981), John Wiley and Sons Ltd., Chichester.

Ward, R.J. and Wood, B.J.

A COMPARISON OF EXPERIMENTAL AND THEORETICALLY DERIVED SENSITIVITY FACTORS FOR XPS

X-ray photoelectron spectroscopy sensitivity factors for elements from lithium to zinc have been determined experimentally using standard compounds. These are found to differ quite considerably from those theoretically calculated using Scofield's photoelectric cross-sections, but are in good agreement with other experimentally derived databases in the literature. It is concluded that Scofield's cross-sections are in error and that a correction factor must be employed if they are to be used for the calculation of XPS sensitivity factors.

Reprinted with permission from Surf. Interface Anal., 18, 679 (1992),John Wiley and Sons Ltd., Chichester.

Werner, W.S.M.

THE ROLE OF THE ATTENUATION PARAMETER IN ELECTRON SPECTROSCOPY

A detailed outline if a recent formalism to calculate teh depth distribution (DDF) in electron spectroscopy in the presence of elastic elestron scattering is presented and discussed. The DDF is intended to provide a quantitative description of the probability of escape of signal electrons and is therefore of crucial importance in quantitative as well as in angular resolved electron spectroscopy. A quantitative comparison of the present formalism with results based on Monte-Carlo calculations reported in the literature shows very good agreement and qualitative predictions given by transport theory are consistent with the formalism. The results suggest that, as far as the effects of elastic scattering are concerned, the attenuation parameter (AP) (which assumes a central role in the formalism, has an unambiguous definition and can be calculated analytically) can be used instead of quantities such as attenuation length and escape depth.

Reprinted with permission from J. Electron Spectrosc. Relat. Phenom., 59, 275 (1992), Elsevier Science Publishers B.V., Amsterdam.

Werner, W.S.M., Gries, W.H. and Stori, H.

ANALYTICAL EXPRESSION DESCRIBING THE ATTENUATION OF AUGER ELECTRONS AND PHOTOELECTRONS IN SOLIDS

The attenuation of Auger electrons and photoelectrons in solids, as described by the so-called depth distribution function (DDF) has been studied by means of a Monte Carlo simulation of electron transport in matter. Elastic scattering plays a significant role in the model. Based on the results of a large number of simulations, an empirically derived analytical expression for the DDF is proposed that includes the effects of elastic scattering. As part of this analytical DDF, a new attenuation parameter (AP) is introduced, which, being a material constant independent of emission angle and layer thickness, can assume a central role in quantitative AES and XPS. The proposed AP as well as the general validity of the results is discussed. Most importantly, a simple relationship between the most significant quantities governing the transport of electrons in solids, i.e. the inelastic

mean free path (IMFP), the total mean free path (TMFP) and the attenuation
parameter, was derived from the results. By a single Monte Carlo simulation,
the AP can be determined. The DDF for different experimental geometries,
film thicknesses, etc. can then be determined analytically.

Reprinted with permission from Surf. Interface Anal., **17**, 693 (1991), John
Wiley and Sons Ltd., Chichester.

Wertheim, G.K.

NEW METHOD FOR BULK QUANTITATIVE ANALYSIS BY ESCA

It is shown that empirical sensitivity factors do not provide a reliable
basis for quantitative chemical analysis by electron spectroscopy for
chemical analysis (ESCA), because the fraction of photoelectric events
accompanied by intrinsic excitations depends on the electronic structure of
the solid. This provides a serious source of error in surface analysis. It
is pointed out that the area of the entire photoemission line, including the
loss tail, is directly proportional to the photoelectric cross-section, and
provides a way of using ESCA for bulk quantitative analysis. A method of
chemical analysis based on this observation is outlined. It has the
advantage of being largely insensitive to surface segregation and surface
contamination.

Reprinted from J. Electron Spectrosc. Relat. Phenom., **50**, 31 (1990), Elsevier
Science Publishers B.V., Amsterdam.

Zalm, P.C.

QUANTITATIVE SPUTTERING

This review discussed experimental techniques used for determining the
variety of observables in ion-bombardment induced erosion of solid surfaces.
Boundary conditions for performing reproducible and accurate sputtering yield
measurements are formulated. To this end, an inventory is made of the
observed systematics in several phenomena accompanying sputtering and of the
more regular exceptions to the general trends, also, a brief treatment of
some theoretical predictions is given, based on the available literature.
Following this extensive survey, the merits and limitations of a number of
measurement methods, tailored to suit one or more different observables, are
discussed. Techniques that seem widely applicable or highly promising are
emphasized.

Reprinted with permission from Surf. Interface Anal., **11**, 1 (1988), John
Wiley and Sons Ltd., Chichester.